A Primer on Alternative Transportation Fuels

Timothy Coffey

Center for Technology and National Security Policy

National Defense University

September 2010

Timothy Coffey served as the D irector of Research of the U. S. Naval Research Laboratory from 1982 to 2001. From 2001 until 2007 he held a jo int appointment as Senior Research Scientist at the University of Maryland a nd as the Edison Chair for Technology at the National Defense University. He retired in 2001 and presently is under contract to the National Defense University as a Distinguished Research Fellow.

Contents

Executive Summary

A review is undertaken of several approaches to produ cing alternative transportation fuels using feedstocks that are under the control of the United States. The objective of the review is to provide the non-specialist reader with a general understanding of the several approaches, how they com pare regarding pro cess energy efficiency, their indiv idual abilities to provide f or national transportation fuel needs, and their as sociated capital costs. It is noted that, in principle, vehi cle missions determine fuel and propulsion plant requirements rather than the other way around. In reality, of course, there is a tradeoff among desired mission capabilities and fuel and propulsion plant technologies.

The review results suggest these conclusions about alternative transportation fuels:

- If necessary, the United States can manufacture the transportation fuels it needs.

- The capital investments needed to ma nufacture fuels beyond petroleum will be substantial, regardless of the particular alternative fuel selected. In this regard, the steam reformation of m ethane (SMR) processes, becau se of their higher efficiencies and substantially lower capital costs, would seem to warrant special attention. The associated fuels are not carbon free or carbon neutral.

- The capital investments associated with the manufacture of renewable—carbon free or carbon neutral—fuels will be especially large.

- Serious commercial investment in altern ative fuels, in contras t to s tandard petroleum-based fuels, will be difficult to obtain as long as low-cost petroleum is available.

True comparative evaluation of tran sportation fuel inextricably requires consideration of the vehicle and mission that the fuel is intended to p ower. A m ission can place significant volume and weight constraints on its combined (power plant plus fuel) system for transportation. Consequently, power de nsity, energy density, and propulsion plant energy efficiency become essential m etrics for comparing var ious alternative transportation fuels. This rev iew commences with a brie f discussion regarding the relationship between m ission requirements and vehicle kinem atics and how these determine the requirements for fuel and propul sion plants. This is followed by a review of the characteristics (e.g., torque, horsepower, weight, volume, and fuel consumption) of several of the most popular engines, where it is shown that the vari ous engines generally are best suited to different m issions. We then focus our attention on fuel approaches where the required feedstocks are within th e direct con trol of the U nited States. The specific topics discussed are: batteries and fuel cells, hydrogen, coal to liquid, natural gas and gas hydrate-derived fuels, bio-derived fuels, CO_2-derived fuels, and oil shale.

For batteries and fuel cells, this r eport starts from a broad discussion of considerations related to the sel ection of the enti re propulsion plant, includi ng fuels and engines. The Ragone chart of specific energy vs. specific pow er is introduced as a comparative means of placing the various propulsi on plants into pe rspective. The vehicle m ission places simultaneous demands on specific power and specific energy. The power plant must be able to m eet these independent req uirements. Hence, the m ission requirements when placed on the Ragone chart must lie within the performance characteristics of the selected technology. Each of the technologies considered occupies a particular area on the chart. The chart shows that, for all the availabl e propulsion plants cons idered, the specific energy of the plant decreases as the specific po wer of the pl ant increases. The internal combustion power plants considered had the most robust capability to support both high specific power and high specific energy. The el ectric power sources considered in the chart include batteries and fuel cells. Importantly, the batteries and fuel cells were shown to have limited specific energy at the required specific power. As a resu lt, there has been a very limited overlap between battery and fu el cell power and energy characteristics and transportation vehicle mission requirements.

The problem with currently available or near -term battery and fuel cell technologies is that either the available energy or the ava ilable power is inadequa te for the vehicle mission. In some cases, where the m ission requires high power for only a sm all fraction of a vehicle's m ission, the introduction of an auxiliary pulse power system becomes viable. In this case th e pulse power system can be "charged" at a low power that is compatible with the m ission average power requirements. For a sm all fraction of the mission the pulse power system will delive r the necess ary high power but low total energy. It is possible that batte ry and fuel cell technology comb ined with electric m otors may progress to the point where this approach is viable for passenger vehicles where high power is typically required for less than ten percent of a vehicle mission. This could be an important development, because passenger veh icles account for abou t half of U.S. oil consumption. However, for high-energy, hi gh-power missions, the chart shows that batteries and fuel cell are disadvantaged re lative to the interna l combustion eng ines considered. It seem s unlikely that the d isadvantage can be ov ercome for high-performance DOD, comm ercial, and industria l missions involving transportation fuels. As a result, inte rnal combustion propulsion plants will likely remain the plants of choice for these m issions for the foreseeable future, and with a co ncomitant essential n eed for "petroleum grade" fuel.

Regarding transportation fuels themselves, a central objective is to provide som e sense of the scale of the various undertakings, such as the capital investm ents that would need to be made to bring the various approaches to the point of providi ng transportation fuel independence and the impact that would resu lt on the various feedstocks. Hence, the alternative fuel approaches considered include: hydr ogen production, creation of synthesis gases from various feedstocks follo wed by a fuel synthesis process, enzym atic production of ethanol, the use of biomass oils for biodie sel production, fuel synthesis using atmospheric carbon dioxide as a feedstock, and the exploitation of oil shale.

In order to undertake a stressful assessm ent of an alternative fu el's viability as a solution to the transportation fuels problem, each alternative fuel was evaluated according to its ability to provide a fuel product with the energy equivalent of the 13.4×10^6 barrels per day (BPD) of petroleum currently used by the U.S. transportation system . In t hat regard, it w as noted that, for the typical passenger vehicle, only about 18% of the fuel energy it co nsumes is actually used for the pr opulsion of the vehicle, with m ost of the remainder lost primarily as waste heat. Since p assenger vehicles account for about two-thirds of the transportation fuel consumed, the energy that is actually required by vehicle kinematics is much less than the number mentioned above. A significant improvement in propulsion plant energy efficiency woul d reduce significantly the am ount of transportation fuel needed thereby reducing—i n a larger sense—the need for alternative fuel.

A comparison of the findings for the variou s alternative f uel processes approaches considered is presented succinctly in table ES1 (next page). This table provides estimates of the process efficiencies and rough order of magnitude (ROM) estim ates of c apital costs associated with producti on of hydrogen and the liquid fu els considered, at a scale needed to produce 13.4×10^6 barrels per day (BPD) of oil e quivalent product. The process energy efficiency was defined as the calorific energy content of the fuel product produced by the process divided by the sum of the cal orific energy conten t of the feedstock consumed, plus any additional energy required to enable the process. The table is ordered by increasing capital costs, excep t for oil shale, which is qualitatively dif ferent from the other entries.

As can be seen from table ES1, there are potentially m any processes by which alternative fuels can be produced with the energy content required by the U.S. transportation system. There is a large spread in the energy efficiencies of the proce sses, ranging from 70% for SMR production of hydr ogen to 18% for gasoline production from atmospheric carbon dioxide (CO_2). Similarly, there is a large spread in the estim ated capital costs associated with the various processes, ranging from a low of about 173 billion dollars for hydrogen production using SMR to a high of about 4 trillion dollars for gasoline production from atmospheric CO_2. It is c lear from the table tha t increasing capital costs tend to correlate with decreas ing energy efficiencies. This is not an unexpected result.

Several of the processes shown in table ES1 can be elim inated for reasons of impracticality towards m eeting the scalab ility goal. Corn- based ethanol and biod iesel have been included simply to show where they fall in the hierarchies of efficiency an d cost. In reality, while both fuels m ay have important niche roles to play, neither of the m is a serious candidate for solving the national transportation fuels problem. The necessary feedstocks are simply not available. It is also recommended that hydrogen be set aside from consideration, even though it has the lowest capital costs when produced by SMR techniques. It could be used as a fuel if it w ere absolutely necessary and could be produced in adequate quantity to meet national transportation fuel needs. If produced by nuclear or solar-powered therm ochemical means or by electrolysis it w ould produce no greenhouse gas CO_2. However, the logistical problem s associated with the cryogenic systems or high-pressure system s required to employ hydrogen as a general-purpose transportation fuel make its use as a general- purpose transportation fuel or as a fuel for most DOD vehicles problematic.

Process	Process energy efficiency	ROM capital cost (billions of dollars)	Comment
Steam reforming of methane (SMR) to hydrogen	70%	170	Commercial process,* doubles NG consumption
Biodiesel	35%	200	Commercial process***
SMR to methanol	60%	250	Commercial process*
SMR to gasoline via methanol	54%	300	Commercial process*
Corn ethanol	46%	400	Commercial process***
Hydrogen by biomass gasification	46%	500	Commercial process**
Hydrogen by coal gasification	44%	500	Commercial process,* doubles coal consumption
Coal to liquid	44%	900	Commercial process,* quadruples coal consumption
Biomass to liquid	47%	900	commercial processes**
Hydrogen by thermochemical	50%	1,000	Process under development*
Lignocellulose ethanol	41%	1,500	Process under development**
Hydrogen by conventional electrolysis	25%	2,000	Commercial electrolyzer, 3'rd gen nuclear reactor*
Atmospheric CO_2 to gasoline	18%	4,000	Commercial processes + 3'rd gen nuclear reactor*
Oil shale surface retort approach	?	1,000	Involves massive mining and disposal
Shell oil shale *in situ* retort approach	Less than 50%	Greater than 500	electric power plant only*

Table ES1. Summary estimates of the process efficiencies and ROM capital costs associated with production of hydrogen and several liquid fuels at a scale to produce 13.4×10^6 BPD Oil equivalent product.

Key: *not renewable but can, in principle, meet the BPD goal, **renewable but available feedstock cannot sustainably meet BPD goal, ***renewable but available feedstock cannot meet BPD goal.)

Among the remaining alternative fuel processes identified here, the SMR processes are found to be the m ost energy efficient and have the lowest capital costs. They can also produce a variety of different fuels incl uding hydrogen, alcohols, and hydrocarbons. A negative aspect is that their use would doubl e the consum ption of natural gas and add substantially to CO_2 production. With regard to the increased consumption of natural gas, the recent advances in gas production using hydraulic fracture of shale and the vast reserves of gas hydrates that are believed to exist m ay play a role. The advances in hydraulic fracture of shale are estim ated to increase the U.S. potentially available natural gas resources by one-third, resulting in about 100 years of natural gas supply at current usage rates. If gas hydrates could be sa fely and econom ically obtained, they would potentially provide a large supply of m ethane thereby allowing the m anufacture of transportation fuels for generations. Unfortunately, at this tim e the viability of gas hydrates as a serious so urce of methane is unknown and it will likely be decades before their viability is quantified. Regarding the production of CO $_2$, if it is dee med inappropriate to release it into the atmosphere, then the f act that it would be produced at large central sites would sim plify its capture. Upon capture it could be processed into additional fuels by m ethods similar to those to be described for fuel production from atmospheric CO_2. This would be an expensive undert aking as can be seen from table ES1. However, it would m inimize CO_2 production during fuels manufacture and also reduce the amount of methane need ed. Ultimately, of course, the CO $_2$ would be released into the atmosphere upon fuel com bustion or reforming. As an alternative, the captured CO_2 could also be sequestered if such schemes are allowed. W hether large-scale sequestration of CO_2 will be viable is currently a m atter of study. It is clear that, while SMR processes could supply the needed tran sportation fuels for an extended period, a number of issues m ust be resolved befo re the SMR approach to alternative fuel production can be properly assessed for implementation.

The next process that shows up as being m ost process energy efficient and least capital intensive is the conversion of coal to liqui d fuels (CTL). This process is considerably more costly than the SMR processes. Nevertheless, it too can produce a variety of fuels, including hydrogen, alcohols and hydrocarbons. Its use would nearly quadruple the consumption of coal and reduce the lifetim e of the U.S. coal supp ly from the current value of 280 years to about 80 years (assumi ng that transportation fuel use did not increase from its current value). The use of CTL would double CO $_2$ production. As with the SMR processes, at considerab le expense that portion of the CO $_2$ that results from the CTL process could be captured at the central production sites and converted to additional transportation fuel, thereby reducing the am ount of coal that m ust be mined. Of course, this approach would increase the dem ands on the supply of uranium in the likely event that nuclear energy would provide the pow er needed to convert the captured CO $_2$ into a transportation fuel. If permitted, here also the captured CO_2 could be sequestered. In this regard, one should keep in m ind that after about 80 years one would have sequestered about half the carbon in the U.S. c oal supply in the for m of gaseous CO $_2$. As with the SMR processes, the CTL process could supply the needed transpo rtation fuels for an extended period. However, a number of issues must be resolved before the CTL approach to alternative fuel production can be properly assessed.

The biomass to liquid fuels (BTL) process comes in at about the same process energy efficiency and capital cost as does the C TL process. The BTL process is generally described as renewable, because the source can be regro wn. However, if it is to be sustainable, then the biom ass harvest m ust be limited by the rate of regrowth of the biomass. Furthermore, there are other uses for the biomass that further lim it the amount of biomass that can be used for alternative fuel production. It has been estim ated that about 1.4 billion tons per year of biomass can be made available, in a sustainable fashion, for the production of alternative fuels. If this biomass were converted into fuel in the BTL process, it could provide for about 35% of th e current transportation fuel needs. Thus, a full solution of the transportation fuels probl em via the BTL approach d oes not seem to be viable. Of course, a 35% solution would be a significant contribution. The capital cost associated with the 35% solution would be a bout 300 billion dollars rather than the 900 billion dollars indicated in table E S1 for a full solution. In this regard, it should be recalled that the SMR processes indicate that a full solution may be available for a capital investment of about 300 billion dollars. In support of the BTL process it is often suggested the BTL approach is CO_2 neutral, because the carbon in biomass com es from the atmosphere and is returned to the at mosphere upon the com bustion of the BTL fuel. This statement is true to the extent that the energy to m ake the biomass comes from the sun. However, to the extent th at fossil fuel is required to grow/harvest the biom ass, the statement is not tru e. As with the o ther processes, a num ber of issues must be resolved before the BTL contribution to alternative fuel production can be properly assessed.

The next process to appear in the efficiency and capital cost sorting is lignocellu losic ethanol production. This process attempts to break down the cellulo se and hemicellulose in biomass into fermentable sugars. This pr ocess is still under developm ent and is m uch more difficult than the well-dev eloped corn et hanol process. However, if successful, it has the advantage that it can acces s a m uch larger feedstock than can the corn ethanol process. In order to function in a su stainable fashion, it would target the sam e 1.4 billion tons of biomass as does the BTL process. Th is amount of biomass used as feedstock for the lignocellulosic ethanol pro cess would yield about 30% of the ethanol required for a full solution. The capital invest ment required is estimated to be about 500 billion dollars. The situation regarding carbon neutrality is the sam e as for the BTL approach; the process has yet to be demonstrated at a s cale where its contribution to alternative fuels can be properly assessed.

The fuel synthesis schem es with the lowest energy efficiency and the highest cap ital cost involve the use of atm ospheric CO_2 as a feedstock. Th e process involves absorbing CO_2 from the atmosphere and recovering it for reaction with hydrogen in order to make methanol. The process could stop with m ethanol, or the m ethanol could be further processed to m ake a higher energy density fu el such as g asoline. The recovery of CO_2 and the p roduction of the necessary hydrogen are energy intensive and would likely involve the use of nuclear reactors. The various steps of the process have been tested and involve well-understood technologies. In principle, the process should be capable of producing the quantities of fuel needed to solve the transp ortation fuel problem. For the case considered herein, this would require about 1600 nuclear reacto rs each providing a thermal power of about three gigawatts. The to tal capital cost for the production plants is estimated to be about four trillion dollars. Sin ce the feedstocks for this process are atmospheric CO_2 and water, and the supplemental pow er is from nuclear energy, the process is CO_2 neutral. However, the uranium c onsumption resulting from the large number of nuclear reactors would be substan tial and would likely deplete the reserves of low-cost uranium ore in a few decades. The long -term viability of such a process would likely involve the introduction of advanced reactors, and ul timately of breeder reactors. Similar approaches could be applied to converting sm okestack CO_2 to fuel where the higher CO_2 density would m ake the collection of CO_2 easier. However, because the capital costs for fuel production are dom inated by the recovery of the absorbed CO_2 and the production of the required hyd rogen, for the sam e amount of fuel the plant capital costs would be si milar to that given a bove for fuel production from atmospheric CO_2. There are clearly m any issues th at need to be resolv ed regarding this approach to alternative transportation fuels.

The final topic considered was that of transportation fuel from oil shale. Oil shale processing is qualitatively different from the schemes discussed above. The organic carbon that would form the basis of fuels is obtained by heating the shale to about 700 $^\circ$F and driving a liquid crude oil from the shale. This heating is performed in a vessel called a retort and the process is called retorting. Th e crude oil obtained is collected and sent to a refinery where it is turned into usable fuels m uch as is done with conventional petroleum. Retorting is done on the surface or *in situ*. It is estimated that U.S. oil shale formations could supply current U.S. transpor tation fuel needs for m ore than 140 years. However, the process for obtaining oil shale crude is very energy intensive. There are no recent studies that p rovide current data rega rding the expected cap ital costs as sociated with producing shale oil crude. Studies done in the 1980s, when scaled to 2005 dollars, suggest that the capital costs associated with producing 13.4×10^6 BPD of oil shale crude by surface retorting w ould be ab out one tr illion dollars. Regarding the cap ital costs associated with the *in situ* approach, in recent years, Shell Oil has been studying an *in situ* retorting approach in which the heat needed to drive out the oil shale crude is provided by electric heaters p laced within the s hale deposits. This approach would avoid having to mine the sh ale. It can b e shown that the el ectrical power s ystem needed to p rovide the heat necessary to produ ce 13.4×10^6 BPD of oil shale crude would itself have a capital cost in excess of 500 billion dollars. Ther e are m any environmental concerns (e.g., ground water contamination) associated with producing oil shale crude.

Finally, it is im portant to note that, on the basis of nati onal security needs, the DOD could argue to use appropriated funds to pay fo r the development of an alternative fuel to supply its two percent of national trans portation fuel usage. There are several methodologies that could supply D OD transportation fuel needs. However, such an undertaking should be approached with great caution. If DOD were to select a schem e that is not viable f or the larger transportation system then DOD will be le ft with a costly proprietary system, will be unable to benef it from competitive f orces in the large r marketplace, and could find itsel f short of fuel in a tim e of national em ergency. It is certainly not clear at this time which is the best alternative fuel approach for DOD and for the nation. It will likely take decades to sort this out. DOD should be a participant in a national effort to clarify the choices from a perspective of m ission requirements—to ensure that these r equirements will be m et—since it cou ld be im pacted substantially by the outcome.

1. Introduction

The Energy Infor mation Agency (EIA) estim ates that th e United States pres ently imports 58% of its oil (petroleum) and th at this will grow to about 68% by 2030. [1]This fact, combined with the recent escalation in wo rld oil prices, has led to renewed interest in alternative fuels that might reduce or eliminate the growing dependence on foreign oil sources. Of particu lar concern are the applic ations that power the U.S. transpo rtation system. Table 1 summarizes U.S. oil usage for 2007.

Fuel Use	Amount (million barrels per day)
Motor Gasoline	9.29
Jet Fuel	1.62
Distillate (highway)	2.36
Distillate (ships)	.147
Distillate (Other)	1.69
Residual Fuel	.72
Other	4.82
Total	20.65

Table 1. U.S. oil usage for 2007 (Source: Annual Energy Outlook 2009 table 20, Energy Information Agency "Sales of distillate fuel oil by category 2009."

It can be se en from table 1 tha t about 65% (13.4 million barrels per day) of U.S. oil consumption relates to transportation fuels. This number is somewhat misleading in that the energy conversion efficiency of the ty pical transportation pl ant is about 18% (see figure A1 of appendix A). Thus , the energy that is actually needed for transportation propulsion is equivalent to about 2.4 million barrels of oil per day (BPD). The re maining 11 million BPD e merges mostly as waste h eat due to th e underlying efficiency of the energy conversion process. In this paper we will use 13.4 million BPD as the go al that alternative fuels must meet. However, it should be kept in m ind that a radical improvement in the energy efficiency of propulsion power plants would markedly change the transportation fuels requirement.

The Department of Defense (DOD) is espe cially dependent on petroleum . Figure 1 summarizes the DOD energy usage for 2005.

US Department of Defense Energy Consumption, Fiscal Year 2005

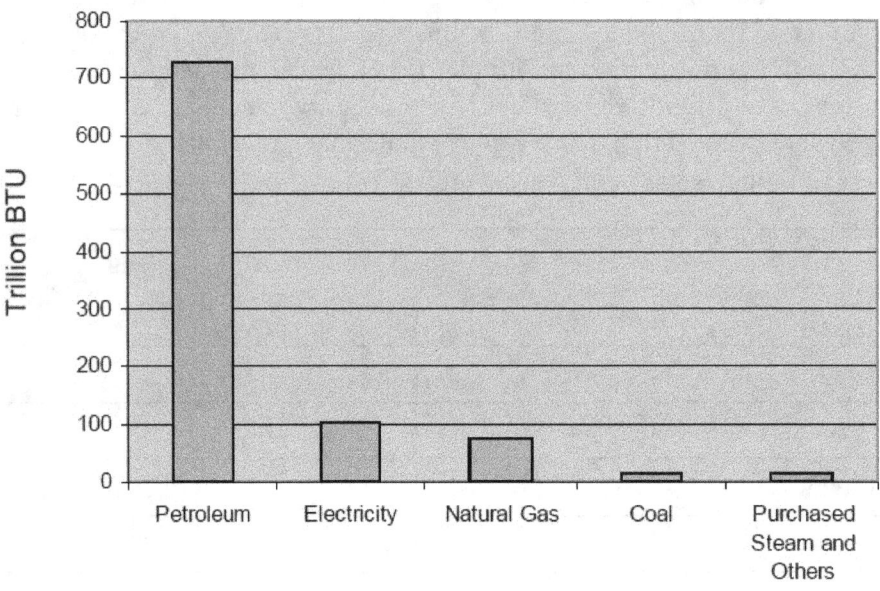

Figure 1. DOD energy usage during 2005 (Source DESC Fact Book FY 08).

It is clear fro m figure 1 that petroleu m is the dominant energy source for DOD. The DOD use of petroleum is primarily associated with transportati on fuels. Since a barrel of oil has an energy content of about 5.8 m illion BTU, it follows from table 1 and figure 1 that DOD is responsible for about 2% of U. S. petroleum consumption. This percentage is sufficiently small that DOD can r easonably expect to have access to the petroleum resources needed to meet its mission in a national em ergency. However, DOD cannot be complacent regarding oil supplies. DOD m ust purchase the oil it uses and is therefore directly impacted by oil prices. Mo re importantly, the missions that DOD will be tasked to undertake are likely to be i mpacted profoundly by international developm ents in the area of energy supplies. Furthermore, DOD has a big stake in the outcome of any attempt to move the nation to an alternative fuel. For example, some of the alternative fuels under consideration would be ill suited for DOD use, as will be discussed in sections 5 through 10. There are other solutions that could meet DOD needs but would not scale to the larger national transportation fuel n eeds. If DOD selected one o f these solutions it would put itself in the position of having to maintain a DOD-unique fuels infrastructure that would not benefit from the potential efficiencies of the larger national fuels infrastructure. Therefore, the large and growing dependen ce on imported oil and the growing national interest in alternative fuels are matters of both economic and military concern. This has led to increased interest within DOD and ot her agencies regarding whether and which alternative fuels could b e produced from U.S. feedstocks on a scale th at could lead to energy independence for the United States.

From a policy and planning pers pective, one of the difficulties with a lternative fuels relates to th e great variety of possibilities that are potentially ava ilable. Each of these possibilities has its own community and literature (some of which goes back 100 years or more). While there is a great deal known about most of the alternative fuels, it is difficult to find discussions that attem pt to place various possibilities within the context of the national transportation fuel proble m and with respect to each othe r. This report will attempt to make some progress in this regard. The objective here is to provide some sense of the scale of the various undertakings, such as the capital investm ents that would need to be made to bring the various approaches to the point of providi ng transportation fuel independence and the impact that would resu lt on the various feedstocks. The paper is intended for the non-specialist for whom a broad understanding of alternative fuels would be helpful. In general, the paper will not de lve into special technical iss ues, such as the best catalyst to em ploy in a par ticular situation. We will as sume such m atters can be resolved, and that it is technica lly feasible to produce the f uel. This report will provide rough order of m agnitude (ROM) estim ates of some of the consequences of m oving beyond technical feasibility to a national solution that pr ovides transportation fue l independence. A detailed discussion of the underlying chem istry and physics can be found in Probstein and Hicks.[2]

Transportation fuels are especially challengi ng, because th ey are typ ically carried by the vehicle that they power. This places significant volume and weight restrictions on the fuels and power plants that can b e considered for transportation purposes. As a result, power density, energy density, and propulsion plant energy e fficiency become essential metrics for comparing various alternative transportation fuels. We will focus our attention on approaches where the required feedstocks are within the direct control of the United States. The specific alternative fuel applications that we w ill discuss are: batteries and fuel cells, hydrogen, coal to liquid, natural gas and gas hydrat e-derived fuels, bio-derived fuels, CO_2-derived fuels, and oil sh ale. The place to begin such a discussion, however, is not with the fuels but with the m ission requirements to which the fuels m ust respond. Section 2 will provid e a brief discussion of some of the ways in which m ission requirements influence fuel and power plant co nsiderations. Section 3 is an overview of the characteristics (e.g., torque, horsepower, weight, volume, and fue l consumption) of several mission-essential engines. C omparative assessment of how the various engines generally are best suited to different m issions, in section 4, incl udes discussion of the electric motor engine and the status of batteri es and fuel cells in th at context. W e then focus our attention on the alternative fuel appr oaches, in sections 5 through 10. In section 11, the review findings are summarized and conclusions are offered.

2. Mission Requirements that Drive System Considerations

While this is not a technical paper, a few basic technical concepts are needed for context. (It is a gratifying demonstration of the power of physics and chemistry that one can make serious progress evaluating alternative fuels by employing just a few simple concepts.) The most elementary concept of importance to this paper is that of mass. By definition, mass relates the motion of an object to a force applied to that object (as in "force equals mass times acceleration"). The subject of transportation fuels is ultimately about the ability to accelerate a vehicle of mass M to some desired speed (velocity) V, within some specified time and to dynamically control the speed so as to accomplish a desired mission. That mission may be as simple as moving a cargo from one point to another or as complex as conducting air combat operations. It is the mission that determines the dynamics required of a specific vehicle. These dynamics are independent (except for their mass) of the propulsion plant and the fuel that powers that plant. In this regard, a simplified dynamical model that relates mission requirements (e.g., required acceleration, required cruising speed, mission duration, vehicle mass and volume) to power plant properties (e.g., horsepower, weight, and volume) is useful. Such a model is presented in appendix A (A Simple Model for Vehicle Kinematics) for the case of land vehicles and some aspects of seagoing vehicles. We will apply this model to several vehicle types in order to gain some insight into the relationship between mission requirements and propulsion plant requirements.

At the level of discussion appropriate to this paper, vehicle dynamics can be broken into two phases, an acceleration phase that moves a vehicle from one velocity to another and a cruising phase that involves motion at a constant velocity. Acceleration capability is usually a stated requirement in the performance specifications for a vehicle. In the case of acceleration from a stopped position, the most important forces are usually the inertial force, the gravitational force, and the rolling force. The rolling force results from energy that is lost during material deformation (e.g., of tires, treads, and pavement) as a vehicle moves. This force is approximately proportional to the weight of the vehicle and, therefore, explicitly involves the gravitational force. Equations (A2) and (A3) of appendix A are helpful in gaining insight into the impact of acceleration requirements on propulsion plant requirements. From these equations it is straightforward to estimate quantities such as horsepower, time required to reach a particular velocity, and the effective energy used during the acceleration process. The effective energy is a kinematic outcome of the mission requirements and is not the energy used by the power plant. The effective energy and the power plant energy are usually related by the efficiency of the propulsion system.

As one might expect, there is often a tradeoff between a desired mission capability and the ability of technology to provide for that capability. In this regard, table 2 provides predictions for several land vehicles of the power and the effective energy that must be provided by the propulsion shaft when the requirement is to accelerate form a stop to 60 mph (27 m/s) in 6 seconds over level asphalt. The table also gives the time to reach 60

4

mph and the energy required to do so when the acceleration is accomplished us ing the horsepower that is actually availab le to each ve hicle. The vehicles have been selected to cover the r ange from personal automobiles to heavy a rmored military vehicles. The vehicle characteristics used to construct ta ble 2 are found in ta ble A.1. The type of propulsion plant that was chosen for the variou s vehicles is indicated by parentheses in column 1. The effectiv e energy th at the chosen power p lants must deliver in order to achieve 60 mph is given in the last two columns.

Vehicle	Vehicle weight (pounds)	Effective horsepower needed for 0 to 60 mph in 6 s	Available horsepower (neglecting driveline loss)	Time(s) to go from 0 to 60 mph with available horsepower	Effective energy (MJ) for 0 to 60 mph in 6 s	Effective energy (MJ) to achieve 60 mph using available horsepower
Tesla Roadster (electric)	2723	215	288	4.4	.48	.48
Honda Accord (gasoline)	3300	252	270	5.76	.58	.58
Jeep GC (gasoline)	4470	363	195	11.5	.79	.83
Hummer (diesel)	6600	520	240	12.5	1.16	1.24
MRAP (diesel)	38,000	2887	330	67	6.2	9.2
Abrams tank (gas turbine)	140,000	11,661	1500	103	24	58

Table 2. For several land vehicles, the horsepower a nd effective energy needed to go from 0 to 60 mph in 6 seconds and also the time to reach 60 mph with the "as-built" horsepower and the effective energy required to do that. The acceleration occurs on a level asphalt surface.

Of the vehicles considered, only the electr ic (Tesla Roadster) and the gasoline (Honda Accord) actually have the horsepower required to travel from 0 to 60 m ph in 6 seconds. The other vehicles are underpowered to m eet this objective. The last two columns compare the energy (in Mega Joules) needed to accomplish the accel eration in the two cases considered. The last two columns are essentially the same for the Roadster through the Hummer. This is because the en ergy needed to ove rcome the rolling force is small compared to the kinetic energy th at was im parted to the veh icles. However, there is a notable difference in the last two columns for the MRAP and the Abram s. This is due to the large rolling force energy loss that results from the weight of the vehicles and the long times that are needed to accelerate thes e vehicles to 60 m ph using the ava ilable horsepower. The third colum n demonstrates the impact on horsepow er if a 6-second acceleration time is imposed as a mission requirement. For the MRAP (which can travel at 60 mph) meeting a 6-second requirement would require a nine-fold increase in the "as-built" engine power, m ass and volume. For the Abrams tank, it would require an eight-fold increase in the as procured engine power, m ass, and volume. The heavy arm or requirements associated with and the volum es available to these vehicles are not compatible with a six-second requirem ent using available propulsion technologies. It is straightforward to show from eq. (A2) that the 1500 horsepower Abram s tank can achieve a speed of about 20 m ph in 6 seconds. (It should be noted here that, while the Abrams tank has the horsepower needed to reac h 60 mph, it is res tricted to speeds less than 42 mph.)

The acceleration calculations summarized above made use of the fact that the desired speed was achieved in distances that were short enough that the drag force had not become important to the ener gy consumption. In the cruise phase, however, the drag force usually plays an essential role. For land vehicles, one can gain som e sense of the relative importance of the drag force and the rolling force b y examining the ratio of drag force to rolling force versus velocity. Figu re 2 displays this ra tio for the veh icles considered in table 2 in the case of travel on asphalt roads.

Figure 2. Ratio of drag force to rolling force vs speed (mph) for several vehicles

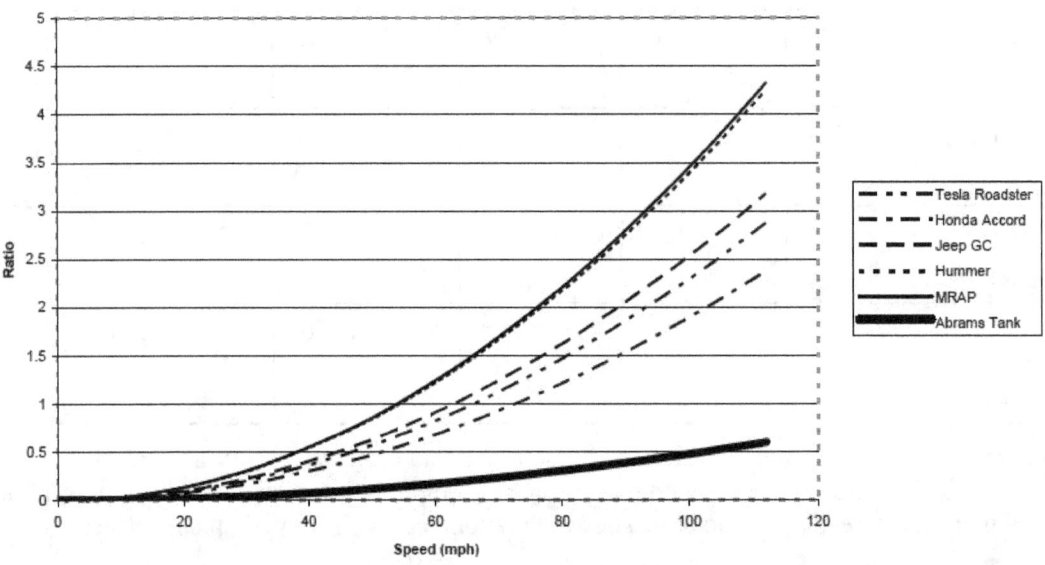

It is clear from figure 2 that th e force that ultimately determines the velocity will var y dependent on the velocity and on the vehicle and its m ission. For example, the drag force is not important for velocities of interest to the Abram s tank. For the other vehicles, the ratio becomes unity for velocities between about 50 mph and 70 m ph. Over their speed ranges, the other vehicles must take into account both the drag force and the rolling force when traveling over an asphalt surface. This could change if the su rface were to change. For example, if the MRAP calculation were don e for travel on soft sand , then the ro lling force coefficient would increase by about an order of m agnitude due to the compressibility of the sand. In that case, the rol ling force would dom inate the drag force for all speeds of interest to the MRAP.

Equation (A5) provides an expression for the horsepower required to maintain a vehicle at a particular speed during the cruising phase. In order to get a sense of that requirement relative to the horsepower requirement to provide for vehicle acceleration, it is informative to consider the ratio of the horsepower that is needed to maintain a vehicle at a particular speed over an asphalt surface to the horsepower available to that vehicle. This is done in figure 2 for the vehicles considered above. To obtain the absolute horsepower one multiplies the numbers in figure 2 by the available horsepower listed in table 2.

Figure 3. Fraction of available horsepower needed to maintain speed vs speed (mph) for several vehicles

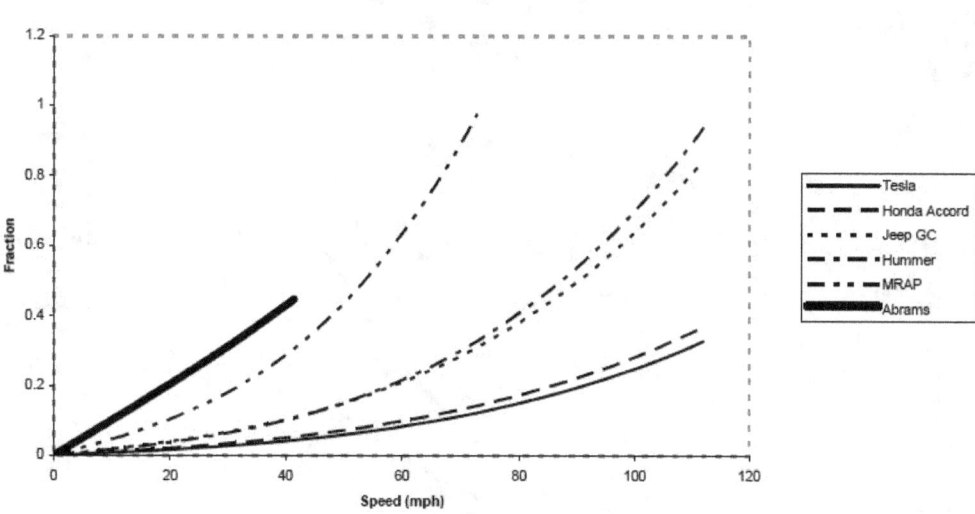

It is seen from figure 3 that, for most land vehicles, the available power exceeds what is needed to maintain the vehicles at a particular speed. For example, the average speed for private automobiles is estimated to be about 45 mph. To maintain that cruising speed, the Honda Accord, as shown in figure 3, requires about 6% of its rated horsepower. At its maximum allowed speed, the Abrams tank, in cruising mode, would require about half of its rated horsepower. The MRAP has a maximum rated speed of 69 mph. At that speed it would be operating at its maximum horsepower. In general, the horsepower of a land vehicle is determined by the acceleration requirements rather than the requirement to maintain a particular speed.

The simplified model described in appendix A can also be used to provide estimates regarding the power and energy requirements of displacement-type seagoing vessels. In that case, the rolling force and the gravity force will be taken as zero. This leaves the inertial force and the drag force as the factors that determine power and energy requirements. The density of water is about one thousand times that of air, and the relevant area in the drag force is the ship wetted area. A simplified model for the wetted area and the drag force for displacement hull surface ships is presented in appendix A.

For purpose of illustration, figure 4 provides es timates of the horsepo wer needed to maintain tanker class ships with lengths between 150 and 400 meters, at a particular speed. A drag coefficient of .004 was taken as being representative of the tanker ships (see Lin et al[5]).

Figure 4. Cargo ship required effective horsepower vs cruise speed in knots (150-400 meter length)

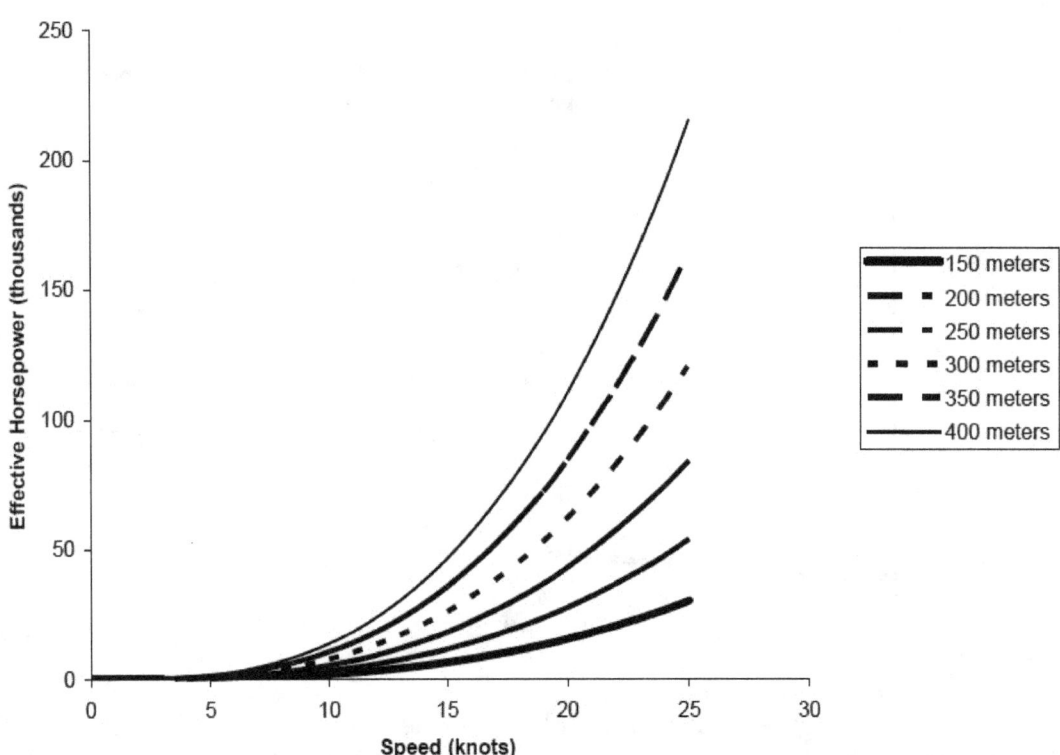

As a point of com parison, the VLCC tanker *Frank A Shrontz*, with a length of about 330 meters and a cruising speed of 16 knot s, has a power plant of about 34,000 horsepower. This is in reasonab le agreement with the predicted value obtained from figure 4 and suggests that the power plants chosen for cargo ships are selected to be close to the power needed to maintain a desired cruising speed. Similar calculation can be done for fast ships (war ships, cruise ships) when the appropriate values of the drag coefficient and wetted surface areas are know n. It is clear th at ships involved in modern commerce, as well as upper end m ilitary vessels and crui se ships, have large power requirem ents. Since these vessels must travel long distances, fuel efficiency can be ex pected to be a significant factor in eng ine selection. The following sections will d iscuss, among other things, factors that influence fuel efficiency.

3. Review of Engine Technologies

The discussion so far has been focused on mission requirements and has been largely independent of the selection of power plants or fuels. Once mission requirements have been determined, then the selection of the engine most appropriate to the mission can be undertaken. While there are a large number of engine technologies available, most engines of interest to transportation fall into the categories of internal combustion (e.g., gasoline engines, diesel engines, and gas turbines), steam turbines, electric motors, and combinations of the above. This section will undertake a brief review of several of these engine technologies and the means used to select among them.

The internal combustion engines and the steam turbines operate by creating a hot gas or vapor and subjecting that gas or vapor to a thermal cycle that converts heat energy into mechanical energy. This conversion may be accomplished by pushing pistons or rotating turbine blades. As a result, the efficiencies of the conversion processes are limited by the laws of thermodynamics. This efficiency is called the thermal efficiency and depends on the thermal cycle that characterizes each engine. The principal thermal cycles used are the Otto Cycle (gasoline engines), the Diesel Cycle, and the Brayton Cycle (gas and steam turbines). These cycles are usually described in terms of a closed curve on a pressure-volume chart. The Otto cycle, for example, involves a closed curve that connects two volumes (e.g., the piston down volume and the piston up volume). Energy is required to compress the fuel air mixture as one goes from the piston down state to the piston up state. A spark is used to ignite the fuel air mixture. This releases chemical energy in the form of heat, which raises the pressure in the compressed gas, thereby creating mechanical energy by pushing the piston so that it travels back to its down state. This mechanical energy is transferred to a flywheel that is used to provide torque to power the wheels and energy to recompress the gas. Heat that is left in the gas after the down stroke is removed, the cylinder is recharged and the cycle is repeated. The thermal efficiency of this process is just the energy supplied by combustion minus the heat that is removed when the cylinder returns to the down state divided by the energy that was supplied by combustion. This efficiency is found to be a function of the ratio of the uncompressed volume to the compressed volume and increases as this ratio increases. Because of the properties of gasoline, the permitted ratio is usually less than 10:1, and the ideal thermal efficiency is about 47%. No thermal cycle is ideal. The non-ideal factors reduce the Otto cycle thermal efficiency to about 37%. The Diesel cycle is slightly more complicated than the Otto cycle in that it operates between a constant volume and a constant pressure. Its thermal efficiency, however, also involves a volume compression ratio, which, because of the nature of the diesel compression and of diesel fuel, can be much greater than the Otto cycle compression ratio. Compression ratios of 20:1 can be achieved. This leads to diesel engines usually being about 30% more efficient than gasoline engines. The ideal Brayton cycle operates between two pressures, and its thermal efficiency can be shown to be determined by the pressure ratio. Pressure ratios up to 40:1 have been used and can lead to ideal thermal efficiencies of about 60%. Those interested in the details of these thermal cycles will easily find descriptions via an Internet search.

While the thermal efficiencies are importa nt, from a mission viewpoint, the principal propulsion plant param eters of interest are torque, thrust , power, fuel consum ption, weight, and volume. We will briefly review these for the engines mentioned above.

If we consider the case where the output power is extracted from a rotating shaft, then of particular interest is torque T, which is the amount of forc e applied tangentially to a circle, so that the load at a particular rpm is measured by the torque being applied by the vehicle driveshaft. Torque is routinely measured as part of characterizing a particu lar engine. The torqu e characteristics differ am ong the various types of engines typically used in transportation and influence the decision regarding what type of engine to use in a particular application. In order to illustrate the torque differences am ong engine technologies, Figure 5 com pares the torque vs. rotation rate for several engine technologies applied to engines of sim ilar size (in each case with th e engine set for maximum power and the curve derived from a commercially available engine). Fo r each engine technology, the torque has been normalized to a maximum value of one in order to cleanly display the various torque characteristics on a single chart.

Figure 5. Normalized torque vs rotation speed

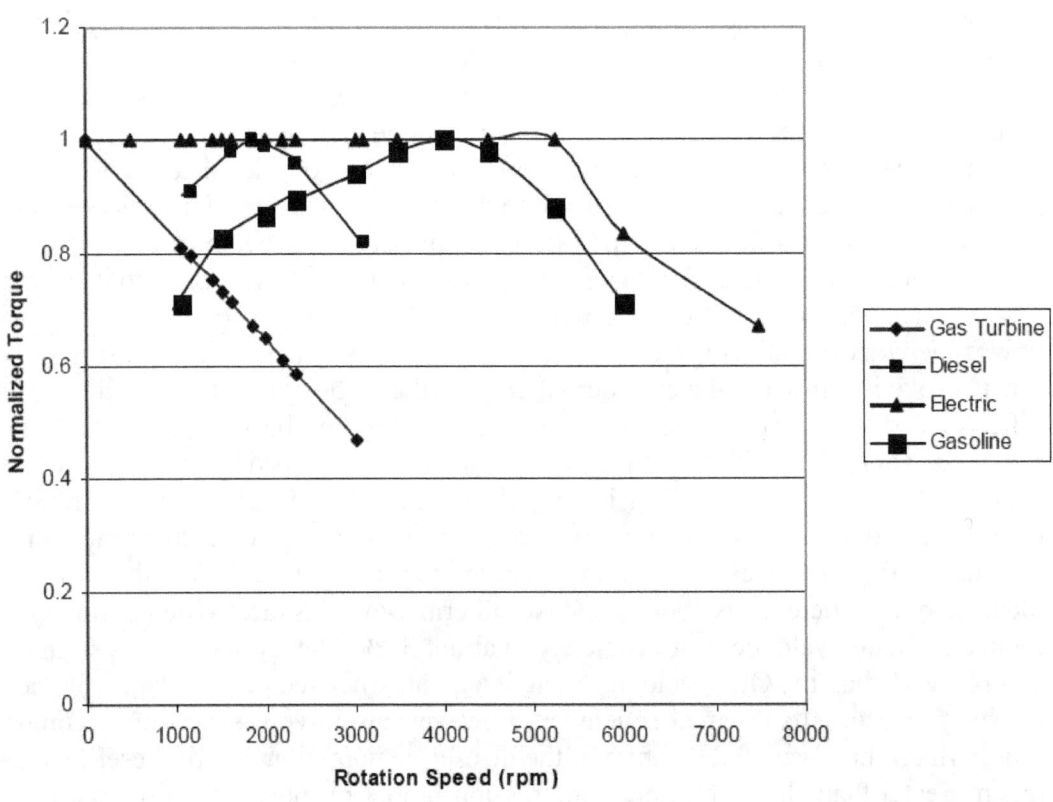

It is evid ent from figure 5 that the vari ous engines have quite different torque characteristics. The electric m otor has close to ideal torqu e characteristics in that its torque is independent of rotation rate until it reaches high rpm. This makes this engine, in principle, broadly applicable. However, as we shall se e later, there are sign ificant challenges in providing the power and energy needed to broadly apply this engine to many transportation applications. The gas turbine engine achieves maximum torque at low rpm with the torque declining as the rpm increases. High torque at low speed is needed for rapid acceleration and for moving large loads. However, for this engine, the maximum torque at low shaft rotation rate is only achievable with the compressor operating at maximum and the output shaft strongly braked. This results in significant fuel consumption at low speeds thereby limiting its application in the low rotation rate arena. In the case shown, the diesel engine provides maximum torque at about 1800 rpm. This torque peak occurs at substantially lower rpm than does the peak for a similar size gasoline engine. This is one of the reasons why diesel engines are used for hauling freight and gasoline engines are not. The gasoline engine curve is broad and provides maximum torque at about 4500 rpm for the case shown. This allows the application of high torque at high speed, which is desirable for passenger cars.

Figure 6. Normalized Horsepower vs Rotation Speed

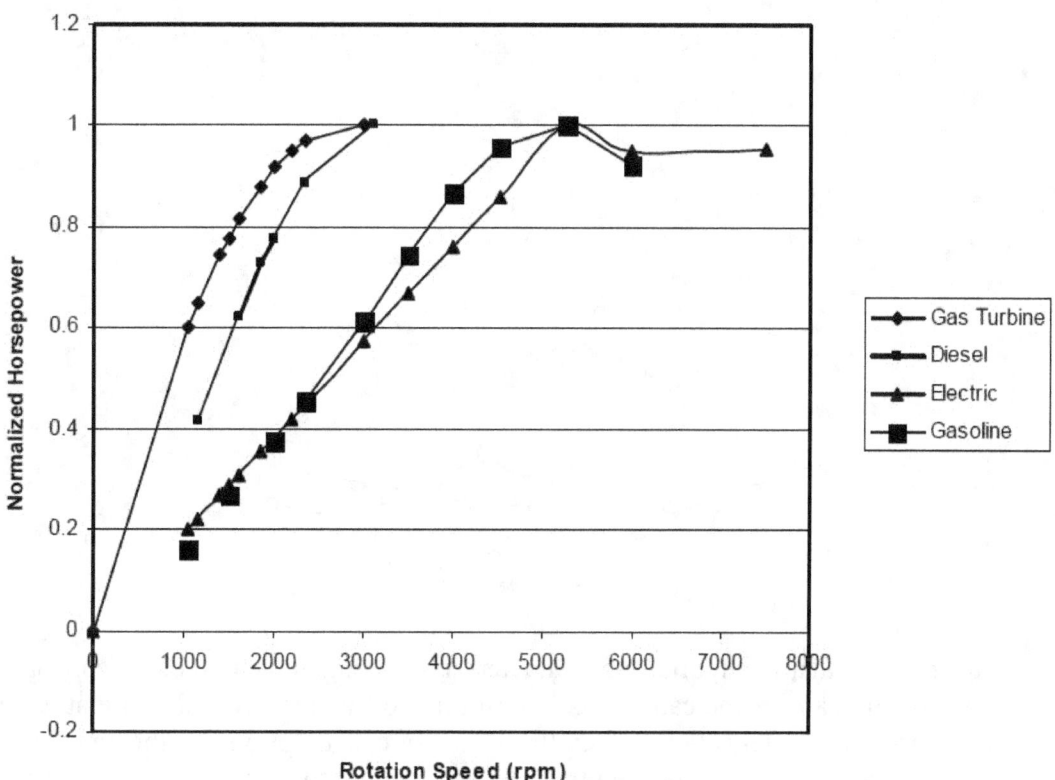

While torque is the parameter that determines the force (and hence the acceleration) that a rotating shaft can apply, the power provided by the shaft determines the rate of fuel consumption needed to accelerate a vehicle and keep it in motion. The power P provided by a rotating shaft is related to the torque by the expression: $P = T \times 2\pi \times f$, where f is the rotation rate of the shaft. Thus, the power is directly obtained from a measurement of the torque. Figure 6 provides the horsepower curves that result from the torque curves shown in figure 5. The power curves have been normalized such that the maximum horsepower for each engine is one.

It can be seen from figure 6 that the horse power curves for the gas tu rbine and for the diesel are similar, although the gas turbine develops its hor sepower more rapidly at low rotation rates. The sim ilarity occurs because the engines used as the basis for the curves were designed for similar applications, namely that of powering light trucks and buses. The curves for the electric engin e and the gaso line engine are also sim ilar, because the engines that for med the basis for th ese curves were designed to power passenger cars. Figures 5 and 6 make the obvious point that the mission of the vehicle strongly influences the selection and design of the propulsion power plant.

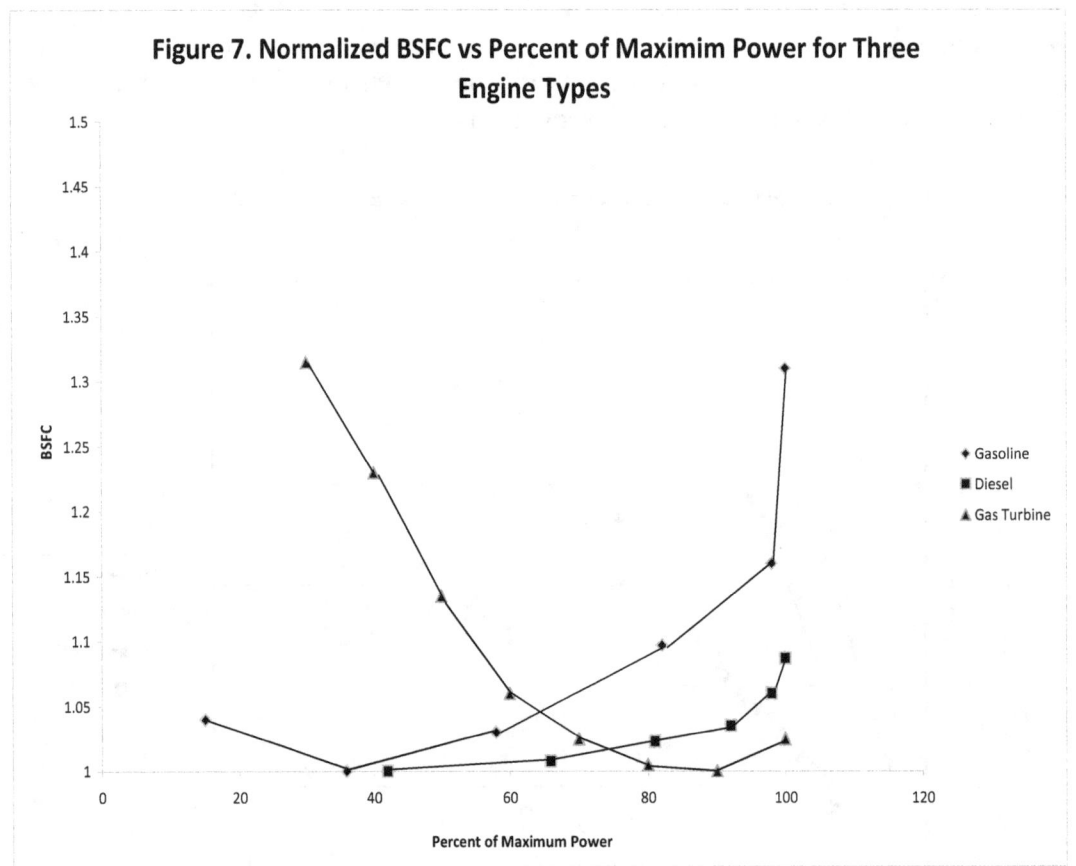

A third factor that is important in characterizing an engine is its fuel consum ption. The energy flow into an eng ine can be easily de termined by measuring the rate at which fuel is consumed. If the fuel is liquid, then th e rate of consumption might be m easured in grams per second. If it is a gas, then the rate of fuel consumption m ight be measured in cubic meters per second. If the energy supply is electric, then the supply current and the supply voltage would be m easured to give the energy input rate directly in Watts. In the case of internal com bustion engines, the chemical energy content of the fuel is released upon combustion. For exam ple, gasoline has a chemical energy content of about 43 MJ/kg, while natural gas has energy content at normal temperature and pressure of about 38 MJ/m^3. Thus, if one m easures the consumption of gasoline in kg/s, or that of natural gas in m^3/s, then it is straightforward to obtain the energy consumption rate in MJ/s. The energy consumption is, of course, specific to th e load that has been applied to the en gine and the rate at which the engine is turning. These can be directly measured. The results of these measurements are usually expressed as th e ratio of the rate of fuel consum ption in grams per second to the power in Watts (J oules per second) being produced by the engine. This ratio is ref erred to as the br ake specific fuel cons umption (BSFC) and is often expressed in gram s per kilowatt hour. Fi gure 7 plots the norm alized BSFC versus the percent of m aximum power for the th ree road vehicle internal combustion engines considered above.

The BSFC in figure 7 is normalized to the minimum BSFC for each en gine (with the engines set for m aximum power). BSFC of un ity, or the p oint where each curve to uches the power axis, is the p ower at which m inimum fuel consumption occurs. The gasoline and diesel engines show a BSFC m inimum in the vicinity of 40% of m aximum power. The gas tu rbine engine shows a minim um at 90% of m aximum power. From a fuel consumption viewpoint, it is clear that thes e engines are best suited for different applications. The gas turbine shows high fuel consumption at low power and is best suited for applications where it will oper ate near m aximum power most of the tim e or where high torque (or thrust) at lo w speed is a requirem ent that overrides high f uel consumption. The gasoline engine shows hi gh fuel consumption when operated near maximum power and is best suited to applic ations where m aximum power is required during only a sm all fraction of its operation. T he diesel en gine demonstrates relatively flat BSFC over m uch of its power range but the BSFC does increase rapidly as it n ears maximum power. W hile it is possible to modify the BSFC curves by various engine modifications, the basic trends shown in figure 7 are intrinsic to the engine types.

When two power plants can m eet the m ission torque, power, and fuel consum ption requirements, then other factors, such as engine size and weight , will dete rmine which engine is selected. This leads to the introduction of engine characteristics such as specific weight (lb/h.p.) and specific volum e (cu.ft./h.p.). Table 3 provides a com parison of several engines including these additional characteristics.

It is clear from table 3 that there is considerable variability among the various engines. For example, among the internal com bustion engines, the m arine diesel represents the heaviest technology and requires the largest inst allation volume. On the other hand, it has among the lowest specific fuel consum ption and the highest energy efficiency. Hence, this technology finds its appl ications to m issions where weight and volum e are not a constraint, but fuel efficiency is of great concern. As a resu lt, the marine diesel currently dominates in the area of commercial ship p ropulsion. At the other extrem e is the gas turbine. This is the lightest technology and requires the least volum e. However, it has rather high specific fuel consum ption and modest energy efficiency. This technology finds applications where weight and volum e are constrained and high power is desired but where fuel efficiency is not a m ajor constraint, or where the eng ine is expected to operate at m aximum power m ost of the tim e. This tech nology is u sed in m ilitary platforms such as tanks and ships and, of course, dom inates the field of a ircraft propulsion, where several variations of the technology (e.g., turbo shaft, low bypass turbo fans, turbo jets) are employed depending on the aircraft mission.

Engine	Specific Weight lb/h.p.	Specific Volume cu.ft./h.p.	Minimum Specific Fuel Consumption lb/h.p.hr.	Engine Energy Efficiency
Low-rpm Marine Diesel 109,000 h.p.	42	0.35	0.26	0.54
Diesel 160 h.p.	7	0.3	0.22	0.35
Gasoline 180 h.p.	2.5	0.05	0.45	0.25
Gas Turbine 1500 h.p.	1.7	0.03	0.45 0.4	
Gas Turbine 57,330 h.p.	0.3	0.03	0.33	0.42
Electric 288 h.p.	.25	0.02	NA 0.9	

Table 3. Specific weight, specific volume, specific fuel consumption, and energy efficiency for several engines.

The electric motor stands out among the several technologies listed in table 3. It is the lightest technology, requires the least volume and has the highest energy efficiency. All things being equal, this would seem to be the best technology for delivering shaft horsepower. However, it will become clear in the following section that the application of electric motors to transportation propulsion is quite challenging. If the challenges can be overcome, the electric motor could have a very large impact on the choice of alternative transportation fuels.

4. Selecting the Power Plant

Section 2 provided a brief overview of how mission requirements translate into dynamical and kinematical requirements. Section 3 provided a brief review of engine technologies that might be compatible with those requirements: diesel, gasoline, gas turbine, and electric. This section will provide a brief review of how to evaluate various power plants within the context of specified missions, and also includes an overview of the current status of batteries and fuel cells in the discussion of the electric motor engine.

Engines must be supplied with fuel (energy) if they are to do anything. Figure 8 provides a simple conceptual drawing of how power might flow in a propulsion power plant.

The "engine power" circle represents the components that provide the power to generate the force that propels the vehicle. The engine power usually conveys via a rotating shaft that generates torque to drive wheels, treads, or propellers or through an exhaust fluid that provides thrust as in a jet engine. The "fuel energy" circle represents the component that provides the primary energy to power the engine. This could be a tank of fuel (e.g., gasoline, diesel, ethanol, and methanol), a pack of batteries, or a fuel cell and its fuel. The primary power supply may not be able to generate the peak power required by the vehicle's mission even though it must be able to provide the average required power. In these cases it is necessary to incorporate a power source that can provide the required peak power. This power source is designated by the "pulse power" circle. The pulse power source may be a battery or may be composed of devices such as electrical capacitors. The intent of the pulse power unit is to provide high power for short times such as when acceleration is needed to climb a hill. The pulse power source will need to be charged or recharged from either the primary fuel or from output power of the engine as indicated by the dashed lines in figure 8. While the figure shows only one engine, it may involve several engines, such as occurs with the combination of internal combustion and electric engines in hybrid drive vehicles. The quantity P_F is the primary power delivered by the fuel. For example, if the engine were an internal combustion engine then P_F would be the energy content per kilogram of fuel times the number of kilograms per second of fuel being supplied to the engine. If the engine were an electric motor, then P_F would be the voltage times the current being supplied by the batteries, electric generator, or fuel cells. The quantity P_{FP} is the primary fuel power that is diverted to charge the (auxiliary) pulse power system. The quantity P_{IN} is the input power to the engine, while P_{OUT} is the output power of the engine. These quantities are related by the efficiency of the engine. The quantity P_{EP} represents engine output that is diverted to charge the pulse power system and P_{PP} is the delivered pulse power. The quantity Pe is the power that is actually provided to propel the vehicle and is determined by mission requirements. The mass appropriate to the power plant is the sum of the engine mass M_e and the fuel mass M_f, where M_f includes the fuel and the pulse power masses.

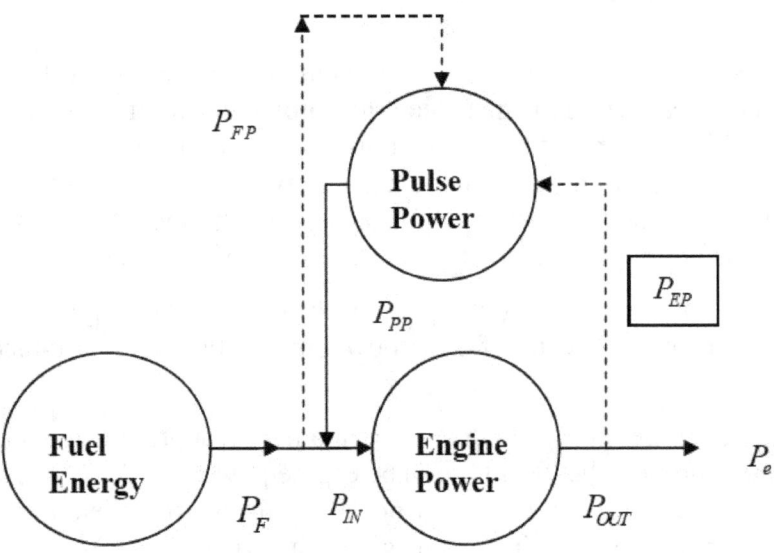

Figure 8. Conceptual outline of power flow within a vehicle power plant.

The energy E_s that can be extracted from the various energy sources is mostly well understood. This energy is supplied to an engine that can turn it into useful work with an efficiency η. It is obvious that the energy E_s is a function of the mission duration and the mission effective power requirements P_e. The fuel mass M_f is often related to the source energy E_s by the expression $E_s = M_f e_s$, where e_s is the specific energy of the fuel. For example, for gasoline $e_s = 12$ Watt hours per gram (Wh/g) of gasoline. This allows the determination of the fuel mass needed to provide the mission energy E_s. The fuel mass is taken to be the sum of the masses of all the components (i.e., fuel, batteries, fuel cell, generator, pulse power) that supply energy to the engine. The energy and power that can be extracted from fuels, batteries, fuel cells, pulse power systems, and generators are determined by detailed measurements that characterize particular fuels, batteries, fuel cells, pulse power systems and generators. These detailed measurements allow one to characterize various power plants in terms of their specific power and specific energy. For a given mission, the specific power SP of the total power plant is defined as

$$SP = \frac{P_e}{M_f + M_e}. \qquad (1)$$

16

Similarly, the specific energy SE of the total power plant is defined as

$$SE = \frac{\eta E_s}{M_f + M_e}. \qquad (2)$$

These quantities are mission dependent. We saw in section 2 that P_e is determined by acceleration and/or speed requirements and by the total mass of the vehicle. For the purpose of discussion, we will assume that the mass of the vehicle is much greater than the sum of the fuel mass and the engine mass. In this case, P_e becomes independent of M_f and M_e. As a result, SP will decrease as M_f increases (i.e., as the mission duration increases). The source energy E_s is taken to be proportional to the fuel mass M_f. Thus, for very long missions, when the fuel mass is much greater than the engine mass, the power plant specific energy will become constant and equal to the specific energy of the fuel source times the conversion efficiency of the engine. The conversion efficiency is a function of the power being supplied to the engine. At the other extreme, when M_f is much less than M_e, the power plant specific energy will scale as M_f / M_e.

Power has units of energy divided by time. Thus, the ratio SE / SP defines a time t where

$$t = \frac{\eta E_S}{P_e}. \qquad (3)$$

The time t provides an estimate of the time to deplete the fuel supply and, hence, of the time between refueling or recharging.

Ordinarily, the mission requirements will determine P_e, $M_f + M_e$ and t. Once these are specified, then SP and SE become determined for the mission. Since only $M_f + M_e$ is specified, there is a tradeoff between M_f and M_e. As an illustration of the outcome of such a specification/tradeoff, table 4 provides estimates of SP, SE for several commercial engines that have been chosen to respond to the specifications indicated. The specifications have been selected to cover the range from passenger cars to massive cargo ships.

The specific powers shown in table 4 are calculated for maximum power. Hence, the discharge times shown are for maximum power. As the power level is decreased below maximum power the discharge time will increase accordingly. For example, the Honda J30 engine operating at 10% of maximum power would have a discharge time of about 8 hours with the fuel mass indicated.

The data shown in table 4 provide some insight into the various power plants. For example, the large, low-rpm diesel RTA96-C is clearly designed for specific energy rather than for specific power. Even if the fuel mass of this power plant were reduced to being much less than the engine mass, the maximum specific power would be only about 40 W/kg. This power plant is designed to economically move large masses long distances. The power plants associated with the two gas turbines (LM6000 and AGT-1500) represent a compromise to provide reasonably high specific power and specific energy as required by their military applications. The diesel plant 6D14-3A has been designed to provide moderate specific power and relatively high specific energy. The gasoline plant J30 has been designed to provide both high specific power and high specific energy. The battery electric power plant has opted for a relatively high specific power in order to compete for the passenger vehicle mission. The low specific energy of the power plant is a result of the low specific energy of the battery technology employed in the power plant and reflects the current state of battery technology.

Engine selected	Specified Max Horse-power	Specified Engine Mass + Fuel Mass	Specified Depletion Time (hours) @Max HP	Resulting Engine Mass (kg)	Resulting Fuel Mass (kg)	Efficiency of Selected Engine	Specific Power (W/kg) @ Max HP	Specific Energy (Wh/kg)
Wartsila-Sulzer RTA96-C Diesel (ship)	109,000	1.2×10^7	735	2×10^6	10^7	.5	6.8	5,000
General Electric LM6000 gas turbine (ship)	57,330	10^6	113	7,818	10^6	.4	42.5	4,800
Honeywell AGT-1500 gas turbine (tank)	1,500	2,800	7	1134	1666	.4	390	2,800
Mitsubishi 6D14-3A Diesel (bus/truck)	160	620	4	500	120	.33	192	767
Honda J30 Gasoline (automobile)	245	146	.8	110	36	.25	1,253	1,000
Tesla battery (electric car)	288	483	.2	33	450	.9	445	99

Table 4. Specific power, specific energy, and depletion or discharge time for several engines with typical mission assignments indicated in parenthesis.

Data of the type shown in table 4 provide a straightforward means for making rough comparisons among power plant technologies as they apply to particular missions. In section 3, it was shown that an electric motor has the most attractive torque, efficiency, and weight characteristics of the shaft horsepower systems considered. Let us now consider the application of electric engine technology to transportation power plants that

18

must carry their own energy. Ragone introduced a chart that plots specific power against specific energy.[4] The chart is now widely used to compare power technologies. Figure 9 provides a "Ragone chart" that summarizes the state of the art for several battery and fuel cell technologies.

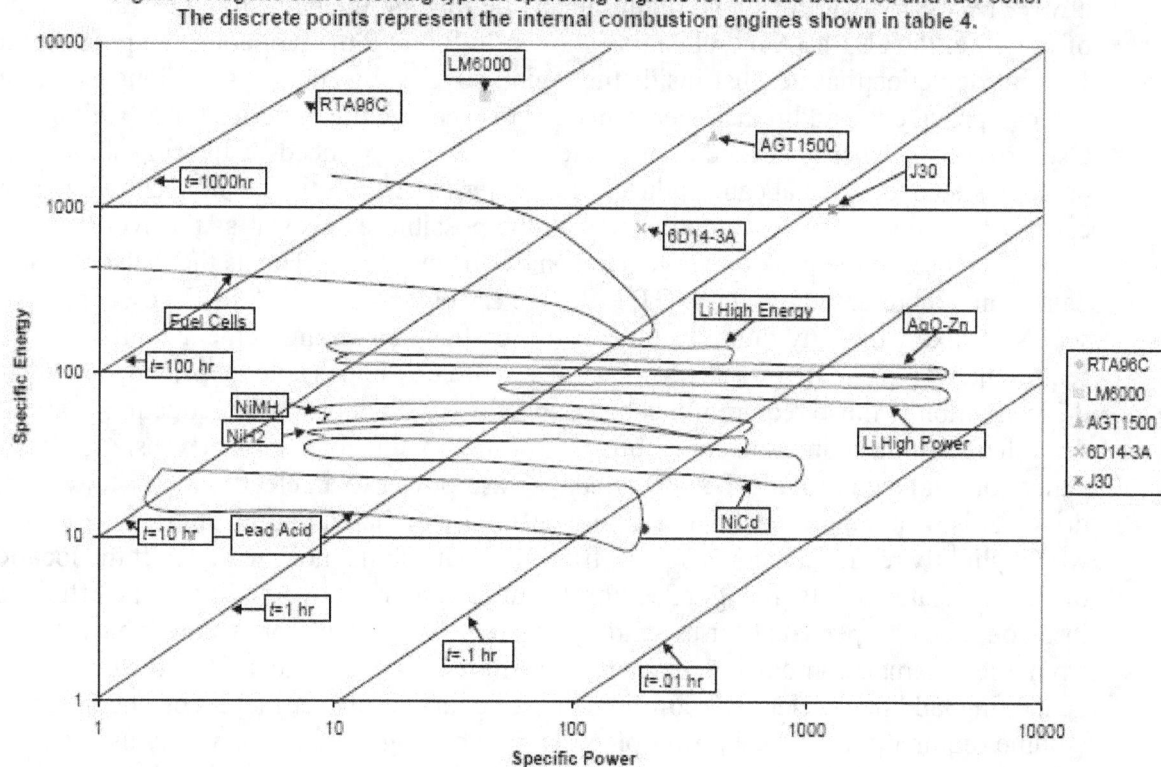

Figure 9. Ragone chart showing typical operating regions for various batteries and fuel cells. The discrete points represent the internal combustion engines shown in table 4.

Sources: Scott,http://ntrs.nasa.gov/archive/nasa/casi.ntrs.nasa.gov/20070021785_ 2007019841.pdf . *Meeting the Energy Needs of Future Warriors,* The National Academies Press, 2004 available at http://www.nap.edu/openbook.php?isbn=0309092612&page=40.

The diagonal lines in figure 9 represent depletion times ranging from 0.01 hours to 1000 hours. The depletion times are indicated for each line. As discussed above, the mission of a vehicle would determine its mass, its required peak power, its average power, the depletion time at peak power, and the mass and volume that can be allocated to the power plant. The ratio of the required peak power to the available power plant mass determines the peak specific power. The intersection of the required peak specific power and the depletion time line determines the required specific energy. This information allows one to obtain a sense of what power plant technologies are available to meet the mission requirements.

The discrete data points on the plot represent the data from table 4 and are evaluated at the mission maximum power requirement. For instance, the J30 Honda gasoline combustion engine can operate at full power for about 1 hour. The internal combustion plant discreet data points shown in figure 9 are representative of DOD requirements for light passenger vehicles, light trucks, land moving equipment, tanks, ships, and aircraft. It

19

is noteworthy that all of these data points   lie outside the regions that are directly accessible by currently available battery and   fuel cell techn ologies. There are b atteries that have the required specific power but not the required specific energy. The fuel cells may meet the specific energy requ irements for average power for the J30 and 6D14-34 power plants but not the specifi c power requirements at maximum power. It is necessary that both requirem ents are m et because they are set by the m ission. However, it is not always necessary that th ese two requirements be met simultaneously. For example, most of the time, the 6D14-3A diesel engine show n in figure 9 m ay operate at a point on the 10-hour depletion that lies just inside the region covered by fuel cells. As long as the fuel cell can provide the additional energy needed to recharge the batteries (or capacitors) and also provide the energy needed to meet the   average power needs, a hybrid solution that combines technology that can pr oduce high power for a short ti me with technologies that can produce low power for long times m ight be possible. However, as the average power becomes closer to the peak power this becomes difficult to do. This is the situation for the data points represented by the AGT 1500, the LM6000, and   RTA96-C power plants. It seems unlikely that any future battery an   d fuel cell arrang ement will m eet the requirements represented by those points. It is, of course, possible to provide an electric drive solution to those requirem ents by connecting (with appropriate reduction gears) the driveshafts of those internal com   bustion power plants to   a properly sized electric generator, and then provide the generated elec tric power to an electric motor for electric drive. This would increase the m ass and vol ume associated with the power plants and would slightly reduce the conve rsion efficiency, but should not greatly shift the location of the associated points in fi gure 9. For the large system s involved here, the efficiency loss could be com pensated for by employing co -power generation, where the waste heat from the in ternal combustion process is captu red to raise   steam for the purpos e of generating additional electric power. This, of course, further increas es the m ass and volume required for the total power plant. Howe ver, cogeneration is widely used and can achieve overall efficiencies of 60%. This approa ch can be quite attrac tive if the mass and volume allocation for the power p lant can acc ommodate it. Such a n electric power solution is actua lly an inte rnal combustion solution r ather than an all-ele ctric solution. The major reason to consider such an appr oach would be the pot ential convenience and flexibility that a large electric power plant might provide (e.g., ship propulsion power and ship onboard power from a common power source).

For shaft horsepower applications, the electric motor is the most attractive in terms of principal propulsion plant parameters, as we saw in section 3. However, the Ragone chart figure 9 sh ows that th e electric motor is currently limited by the sp ecific energy and specific power characteristics of available b attery and fuel cell sy stems. There are ongoing battery developm ents that could im prove the viability of electric m   otors for transportation. One example of this is the lithiu m air battery.[5] This battery chem istry is currently under study, and Ragone charts are not yet availabl e. Based on the theoretical maximum energy density of the lithium air ch emistry and the industrial experience that relates theoretical energy density to energy density that can be achieved in practice, expectations are that a primary lithium air cell c ould achieve an energy density of 1700 Wh/kg. If this energy density could be achieved in a recharg eable lithium-air cell, and if the discharge characteristics allow a specific power in the 1000 W /kg range, then this hypothetical lithium-air cell would approach   the range now occupied by in   ternal

20

combustion engines. In order to gain some insight regarding how this hypothetical battery would impact the matter of transportation fuels, it is interesting to compare the depletion times shown in table 4 with those that would be predicted when the sum of the engine mass and the fuel mass is that shown in table 4 but the power plant is composed of an electric motor and the hypothetical lithium air battery that has an energy density of 1700 Wh/kg and is able to meet the specific power requirements shown in table 4. This comparison is done in table 5.

Engine	Measured Depletion Time (hrs) for Several Engines, from Table 4	Hypothetical Depletion Time (hrs) Using Hypothetical Li-Air Battery and Electric Motor
Tesla (electric)	0.2	3.15
J30A (gasoline)	0.8	0.95
6D14-3A (diesel, land)	4	7
AGT1500 (gas turbine, land)	7	3
LM 6000 (gas turbine, ship)	113	21
RTA96-C (diesel, ship)	735	211

Table 5. Comparison of depletion times from table 4 with those predicted for a replacement of the power plant with a hypothetical Li-Air battery and an electric motor.

In this table, the Tesla electric motor is used for both the replacement of the Tesla power plant and the J30A power plant. The 6D14-3A and AGT1500 engines are replaced by industrial quality electric motors that have specific weights of one pound per horsepower. The LM 6000 and RTA96-C engines are replaced by electric motors that have specific weights of 15 pounds per horse power, which is representative of similar sized electric motors used to power the RMS Queen Elizabeth II.[6]

The hypothetical Li-Air battery considerably increases the depletion time for the Tesla power plant. The J30A replacement plant provides a slightly increased depletion time. Since both the Tesla and J30A power plants are expected to spend most of their operating time at about 10% of maximum power, it seems likely that a day's driving (about 400 miles) could be accomplished with the hypothetical Li-Air battery power plant. This would make overnight recharging of the batteries reasonable at about a 10 KW charging rate. The 6D14-3A power plant replacement provides about a 70% increase in the depletion time resulting in a 7-hour depletion time at maximum power. If one assumes that the application of this power plant will require that it operate on average at half its maximum power, then the time between recharging would be about 14 hours. A 10-hour recharge time would require a 100 KW charging system. This arrangement might be viable for commercial applications (bus, light truck, etc) where recharge periods could be scheduled. However, it is unlikely to be suitable for military applications where vehicles must be available on demand and recharge time requirements are measured in minutes. In the cases of the use of Li-Air batteries and electric motors to replace the AGT 1500,

21

LM6000 and RTA96-C power plants, the replacement power plant depletion times do not meet the mission requirements.

From the above discussion one can conclude that, if the hypothetical, rechargeable Li-Air battery (SP = 1000 W /kg, SE = 1700 Wh/ kg) were realized and found to be economical, durable and safe enough for m anned vehicular use, it would enable the replacement of the internal com bustion-powered passenger vehicle with an all electric vehicle that could travel 400–500 miles on a single charge. It is inte resting to speculate on how such a development would impact the ma tter of transportation fuels. Table 1 lists the 2007 U.S. usage of petro leum and indicates that about 13.4 million barrels per day is used for transportation purposes. A bout 70% of this usage can be attr ibuted to passenger vehicles, which appear to be am enable to being powered by the hypothetical L i-Air battery. Thus, if the hyp othetical, rechargeable Li-Air battery were realized and found to be economical, durable, and safe, 70% of the need for conventional transportation fuel could be elim inated. This outcom e would have a very large im pact on alternative transportation fuels. The burden of fueling vehicles now powered by gasoline would pass to the nation's electric grid. Since passenger vehicles travel about five billion m iles per day, the burden on the grid would be about 10^{11} Watts corresponding to about 20% of current average electric power consum ption of 5×10^{11} Watts. The to tal U.S. elec tric generating capacity is about 10^{12} Watts. Thus, it m ay be possible that this increase could be absorbed by the existing generating capac ity, in which case there would be no power plant construction required to transition passenger vehicles to electric drive on a la rge scale. However, becaus e peak electric pow er requirements can be about 80% of the maximum generating capacity, it w ould be prudent to plan on adding the additional capacity needed to accommodate a large-scale move to electric passenger vehicles.

The technology options for provisio n of this a dditional electrical capacity are likely to be broader than those associated with the production of a hydrocarbon or alcohol replacement of gasoline. For exa mple, the a dditional capacity could be provid ed by doubling the Nation's current nuc lear power electrical genera ting capacity or adding additional coal or natural gas el ectrical power plants.. There is a great deal of uncertainty regarding the capital cost associated with constructing new nuclear power plants. A unit cost of about $1500/KW$_e$ is at the lower end of estimates that are now appearing in the literature. This unit price is also representati ve of coal-fired power plants. At th is unit price, the capital cost of adding 10^{11} W would be about 150 billion dollars. There would, of course, be additional capital inv estments needed to put in place th e infrastructure that would produce and recycle the batt eries and the infrastructure associated with recharging the electric vehicles. The total capital inv estment, however, is likely to be com parable or less than th at associated with putting in pl ace a synthetic fuel production infrastru cture required to replace petroleum -derived gasoline. In ligh t of this, it would seem that a resolution of the viability to transportation systems of Li-Air-like battery chemistries will be important to de termining the f uture direction for alternative transportation fuel initiatives.

Central questions are: can Li-Air battery technology provide needed acceleration, and, is it viable for heavy vehicles? At this time it is not clear that the Li-Air chemistries will be viable for transportation system s. For ex ample, the high specific en ergies that h ave been reported for Li-Air batteries involve low discharge currents, where the disch arge times are measured in hundreds of hours. W hile the published data is lim ited, it is clear that the energy that can be ex tracted from the batteries d ecreases rapidly with increasing discharge currents, which would be necessary for rapid vehicular acceleration. This is the same phenomenon that is evident for batteries and fuel cells in the Ragone chart given in figure 9. It is possible and ev en likely th at the Li-Air ba ttery will prove to be a high specific energy device but a low-to-moderate specific power device. Furthermore, even if the chemistries do enable large-s cale use for passenger vehicles, it is unlikely that they will enable battery-powered heavy trucks, jet aircraft, armored vehicles, tanks, and large ships.

Missions drive the need for heavy trucks and armored vehicles, and the consequent need for combustion engines—and hence the need for liqu id fuels that have the attractive features of petroleum-based fuel products. It is, therefore, prudent to continue to investigate alternative means to the equivalent of conventional fuels, especially those that are not likely to be supplanted by all-el ectric propulsion. The following sections will provide an overview of several approaches in this regard. W e start with hydrogen, followed by coal to liquid, natural gas and ga s hydrate-derived fuels, bio-derived fuels, CO_2-derived fuels and, finally, oil shale.

5. The Role of Hydrogen

This section will p rovide a brief discussi on of the ro le of hydrogen in alternativ e transportation fuels. Hydrogen is broadly important to the subje ct of alternative transportation fuels, so is conside red first in this review. It may be a constitu ent of the fuel (e.g., hydrocarbons or alcohols), it m ay play a role in battery chem istry or fuel cell chemistry, or it m ay be the fuel itself (e .g., a hydrogen-fueled internal com bustion engine). The prospects for hydrogen as a fuel for DOD were discusse d by Coffey et al. [7] This section will summarize the relevant findings from that paper.

To begin that discussion it is helpful to put hydrogen into perspective relative to several other fuels. It is clea r from eq. (A1) that the fuel m ass that a vehicle m ust carry directly impacts on the vehi cle dynamics. Also, the vo lume constraints on transportation vehicles place significant lim its on the am ount of fuel that a vehicle can carry. Thus, as discussed earlier, energy per unit m ass and energy per unit volume are important metrics for judging the viability of various tran sportation fuels. Since the m ost familiar transportation fuel is gasoline, it is helpful to have an understanding of energy content of various alternative energy sources of interest relative to the energy con tent of gasoline. This is p rovided in tab le 6 f or several energy sources. If one wishes to estim ate real energy densities rather than relative energy densities, one can take the specific energy of gasoline to be about 12,000 Wh/kg, or about 9 Wh/cc. Multiplication of columns two and three of table 6 by the appropria te specific energy of gasoline will prov ide estimates of the specific energies of the various energy sources listed.

It should be clear from table 6 why the li quid hydrocarbon fuels (particularly gasoline and diesel) have been dom inant for the past 100 years. W ith the exception of hydrogen, they provide the highest energy content by eith er measure and are easily stored at roo m temperature and pressure. W ere it not for growing concerns rega rding their continued easy availability and their potential adve rse environmental impact, hydrocarbon fuels would remain the obvious transportation fuels of choice. The liquid alcohol fuels ethanol and methanol have the next highest energy content by either m easure (again, except for hydrogen). The hydrocarbon gas methane, when pressurized to 10,000 psi, approaches its liquid energy densities and approaches 50% of the volumetric energy density of gasoline. The listed batte ries have the low est energy content by either m easure (except for hydrogen @ 3000 psi) of any of the fuels listed.

Regarding hydrogen, it can be seen from table 6 that hydrogen has the highest energy content per unit mass of any of the listed fuel s. However, for transportation fuels, energy content per unit volum e is often the m ore important parameter because of volume constraints. By this m etric (with the excep tion of batteries), hydrogen has the lowest energy content of the listed fuels. F urthermore, to obtain hydroge n's highest volum etric energy content it m ust be cooled to $-253\,C^{\circ}$ or pressurized to over 10,000 PSI. These properties of hydrogen m ake it problem atic as a fuel, especially for m ost high-performance DOD platforms.

Energy Source	Energy per unit mass	Energy per unit volume	Temperature $^\circ$C	Chemical Formula
Liquid Gasoline	1.0	1.0	25	$<C_8 H_{18}>$
Liquid Diesel	.97	1.1	25	$<C_{12} H_{23}>$
Liquid propane (@ 90psi)	1.0	.86	25	$C_3 H_8$
Liquid Methane	1.3	.75	-162	CH_4
Liquid Ethanol	.61	.69	25	$C_2 H_5 OH$
Liquid Methanol	.44	.51	25	$CH_3 OH$
Liquid hydrogen	2.6	.27	- 253	H_2
Methane Gas (@ 3000 psi)	1.3	.37	25	CH_4
Methane Gas (@ 10,000 psi)	1.3	.5	25	CH_4
Hydrogen Gas (@ 3000 psi)	2.6	.06	25	H_2
Hydrogen Gas (@ 10,000 psi)	2.6	.2	25	H_2
Lead Acid Battery	.003	.006	25	NA
Li Ion Battery	.013	.026	25	NA
Li Polymer Battery	.01	.029	25	NA
Hypothetical Li-Air battery	.14	.1	25	NA
NiMH Battery	.004	.02	25	NA

Table 6. Energy content and chemical composition of several energy sources referenced to gasoline. The bracket < > indicates the average chemical formula. (Source: modified from Coffey et al.[7])

Hydrogen is the simplest element and can be considered to exist in unlimited supply on earth. When combined with oxygen in a fuel cell it produces electricity (and water) that can be used to power vehicles. The range of specific energy and specific power of typical fuel cells is summarized in figure 9. If pure hydrogen is stored on the vehicle, a fuel cell can convert it to electricity with about 80% efficiency. This is conceptually a straightforward approach and eliminates CO_2 production on the vehicle, but it requires

25

that hydrogen adequate to the vehicle's m ission be stored on the vehicle at high pressure or at cryogenic temperatures. It is also po ssible to produce hydrogen onboard the vehicle by using refor ming techniques to rem ove the hydrogen from fuels such as m ethanol, ethanol, gasoline, and diesel. This conversi on can be done with a bout 30–40% efficiency and requires that fuel adequate to the vehicl e mission be stored onboard the vehicle. This approach eliminates the complexity of the hydrogen storage system but replaces it w ith the complexity of a reform er and a hydr ogen purification system and produces CO_2 as a by product of the reform ing process. It seem s likely, based on the c onsiderations of the previous section, that a hydrogen fuel cell combined with an a ppropriate pulse power system could, in principle, meet the speci fic power and specific energy requirements for passenger vehicles. It is unlik ely, however, th at such an arrangement could meet the requirements for most high-performance DOD or industrial platforms.

Hydrogen has the property that, when com busted with oxygen, it produces only water thereby eliminating the CO_2 associated with combusting hydrocarbon or alcohol fuels on vehicles. When combusted with air, it does produce nitrous oxides, but properly designed engines can solve this problem at the cost of producing engines that are som ewhat larger than gasoline engines of the sam e power. The design of hydrogen-based internal combustion engines goes back to the early 1800s a nd can be considered to be a solved problem. Hydrogen internal com bustion engines can most certainly be designed to m eet civilian and military transportation requirements. The hydrogen could be supplied fro m a cryogenic or high-pressure tank or from reformed fuels. This approach is lim ited by the thermodynamic efficiencies of internal combustion and has the sam e fuel storage/conversion problems associated with the fuel cell s mentioned above. It is not clear that hydrogen as a combustion fuel will prove to be satisfactory or competitive with other alternatives. The cryogen ic and high- pressure requirements will be a great impediment to its adoption. This is especially true for DOD applications.

There are, o f course, special applications where hydrogen will be the preferred fuel. These include situations where the high gas velocities available from hydrogen are needed (e.g., hypersonic aircra ft) or where fuel m ass is more important than fuel volume (certain space app lications). In general, however, it seem s unlikely that hydrogen will prevail as a preferred general purpose or military fuel. This matter has been the subject of extensive debate and will be ultimately resolved by the marketplace.

Regardless of how the debate work s out, hydro gen will play an im portant role in the larger effort to develop econom ically viable and environm entally acceptable alternative fuels. This is sim ply because hydro gen is a very important ingredient for these fu els. However, in order to use hydrogen one m ust obtain it. In general, even though hydrogen is plentiful, it does not exist as a free entity but rather as part of chem ical compounds (e.g., as a com ponent of water, natural ga s, and biom ass). If one envisions using hydrogen as a fuel or in the m anufacture of alternative fuels, it is important to gain some insight regarding the costs a ssociated with producing hydrogen on a scale relevant to the U.S. transportation fuels problem. In this regard, it is helpful to examine the capital costs associated with producing an am ount of hydr ogen equivalent in energy content to that actually needed to p ower the U.S. transportation system (i.e., 2.4×10^6 BPD of oil equivalent, assuming 100% energy conversio n efficiency). Hydrogen has an energy content of about 142 MJ/kg, while a barrel of oil has an energy content of about 6.1×10^3

MJ. Thus, the hydrogen production corresponding to the actual energy needed to power the U.S. transportation is about 10^5 tons per day (TPD). The actual hydrogen needed would depend on the true energy conversion efficiency. For example, if hydrogen were used directly as a fuel and combusted in internal combustion engines with a net conversion efficiency of 18%, then 5.5×10^5 TPD of hydrogen would be required. If the hydrogen were used as a fuel to power PEM fuel cells with 80% conversion efficiencies coupled to electric motors with 90% efficiencies, then about 1.4×10^5 TPD would be needed. If hydrogen were employed in a process to synthesize some alternative fuel, then the amount needed would depend on that process. However, the amount needed would likely be appropriately measured in units of 10^5 TPD. We will, therefore, use that as the appropriate metric for the discussion regarding the production of hydrogen.

There are several well-known methods for producing hydrogen that involve the commercially proven technologies of reforming gases, gasification, and electrolysis. A discussion of these technologies in the context of hydrogen production has been published by Simbeck and Chang.[8] Other promising approaches have been studied but have not been proved commercially viable. An example is the use of thermochemical cycles for splitting water to produce hydrogen.[9]

In order to gain some insight into the costs associated with the various schemes, table 7 provides rough order of magnitude estimates (ROM) of the capital costs associated with the construction of hydrogen production capability using several of these technologies. The estimates are based on published data.

The first three technologies (SM, BG, CG) are well understood, and the cost estimates are based on equipment that has been used commercially for many years. The SI-MHR, or Sulfur-Iodine (SI) cycle that uses heat from a modular helium reactor (MHR) for the necessary thermal energy, is a proposed system that has been well studied but for which the major components are under development. The electrolysis technology for the CE-NR system is well understood, but the unit cost used for the nuclear reactor is at the low end of estimates appearing in the current literature. Thus, the SI-MHR and CE-NR estimates are likely to be optimistic. It is clear from table 7 that by far the lowest capital costs for producing hydrogen are associated with steam reforming of methane (SMR). As a result, the SMR process produces most (95%) of the hydrogen used in the United States today.

The production of CO_2 is significant for several of the schemes shown in table 7. For example, hydrogen production by current industrial SMR processes produces about 9.5 tons of CO_2 for each ton of hydrogen. Hence, SMR production of 100,000 TPD hydrogen will result in about 950,000 TPD of CO_2. Conventional coal gasification schemes produce about 20 tons of CO_2 for each ton of hydrogen produced, resulting in 2×10^6 TPD of CO_2 per 100,000 TPD of hydrogen. Biomass gasification is somewhat more complicated regarding CO_2 production, because the carbon that is gasified to produce hydrogen originated in the atmosphere. In that sense, the process is CO_2 neutral. There may, however, have been CO_2 produced in the process of growing the biomass that would need to be taken into account regarding the CO_2 balance. The SI-MHR and CE-NR schemes would produce no CO_2.

Hydrogen production technology	ROM of capital costs for 100,000 TPD capability of hydrogen production (billions of dollars)	Basis of capital cost estimates	Process Energy efficiency
Steam reforming of methane (SMR)	31	Mintz et al.[10]	70%
Biomass gasification (BG)	92	Spath et al.[11]	46%
Coal gasification (CG)	94	Buchanan et al.[12]	44%
Sulfur Iodide Cycle powered by high temperature nuclear reactor (SI-MHR)	194	Schultz et al.[9]	50%
Conventional electrolysis of water powered by a nuclear reactor (CE-NR)	360	Ivy for CE.[13] $1500/KW$_e$ for NR	25%

Table 7. ROM of Capital Costs Associated With Producing Hydrogen Equivalent of the Energy Actually Needed to Power the U.S. Transportation System.

If the matter of what to do with the CO_2 produced by the above processes can be satisfactorily resolved, then the SMR process would likely remain the preferred means for producing hydrogen on a large scale, as long as methane gas is readily and economically available. The recent advances in the use of hydraulic fracture for the production of natural gas from shale may help in the near term. In the long term, exploitation of gas hydrates could become important to ensure the continuing supply of methane needed for SMR. If the CO_2 disposal problem cannot be resolved, and large amounts of hydrogen are needed then biomass gasification, thermochemical cycles, electrolysis, or some other method of production will need to be considered. There will be feedstock issues (e.g., availability of required biomass, uranium, and water) that may limit the applicability of each of these approaches. Some of these issues will become apparent in the following sections.

According to table 7, the capital costs for producing a quantity of hydrogen gas with the energy equivalent of 2.4 million barrels of oil per day are estimated to range from about 31 billion dollars (the SMR process) to about 360 billion dollars (the conventional electrolysis process using nuclear electric power). These costs do not include capital costs associated with feedstock acquisition and delivery, hydrogen liquefaction (if used), storage and distribution of the hydrogen or if necessary the disposition of the CO_2 produced.

The capital costs provide a sense of the scale of the undertaking and the funds that need to be invested or borrowed in order to proceed. They do not, however, translate directly into the cost of the produced product. The cost of the product produced depends on factors such as operating costs, feedstock pr ices, interest rates, payback time for the borrowed money, and tax rates. These factors can be quite variable and are difficult to project into the future with any degree of certainty. This paper will, therefore, not attempt to estimate product prices for fuels produced by different approaches.

We now t urn to discussion of several methods of producing alternative liquid transportation fuels. For m ost of the altern ative fuels considered in sections 6–10, the production of hydrogen is a dom inant cost. Moreover, in m any cases the hydrogen production schemes discussed in this section are integral to alternative fuel synthesis.

6. Liquid Transportation Fuels from Reforming and Gasification

Today, the primary feedstock for transportatio n fuels is a natural resou rce, petroleum. One of the reasons for looking beyond petrol eum is the growing concern regarding its future availability. It is widely believed that petroleum and other fossil fuels were created over millions of years from the oils (lipids) of plants and animals. This proc ess was highly inefficient. For example, Dukes[14] estimates that, at current consumption rates, one year's worth of energy derived from fossil fuel required 10^{26} Joules of solar energy to produce the plant and animal life needed to make the fossil fuel. That corresponds to an energy efficiency of about 3×10^{-6}. The prac tical consequence of this is that, today, petroleum is being extra cted at a m uch faster rate than it is being crea ted. As a result, petroleum will ultimately be depleted to the point where it is no longer a viable feedstock for fuels. There is general agreement that this will happen, but considerable disagreement as to when it will happen. Perhaps the m ost referenced work in this area is that of M. K. Hubbert, who in 1956 predicted that U. S. oil production would peak in 1970, [15] and in 1969 predicted that world oil production would peak in 2000. Hubbert's prediction that U.S. oil production would peak in 1970 turned out to be correct. His prediction that world oil production would peak in 2000 does not appear to have been correct, and there remains considerable speculation as to when world o il production will peak. Ther e is, however, a growing consensus that it would be prudent to look beyond conventional petroleum.

Carbon-based transportation fuels (e.g., hydrocarbons and alcohols) have m any attractive features. They also hav e the pr operty that, when com busted with air or reformed, they produce CO_2 as a byproduct. In most cases, the synthesis of carbon-based fuels will also produce CO_2. In spite of this latter propert y, it is necessary to have an understanding of how one can synthesize hydrocarbon and alcohol fuels, because they are likely to remain the dominant transportation fuel.

The hydrocarbons are just com binations of carbon and hydrogen, and the alcohols are combinations of carbon, hydrogen, and oxygen. The synthesis of carbon-based fuels is a mater of carbon chemistry about which a great deal is known. In order to undertake such chemistry one m ust have sufficient f eedstocks of carbon, hydrogen, and oxygen. Hydrogen and oxygen are, of course, constituen ts of water and can be considered, in a global sense, to be in unlimited supply.

Where do we get the carbon we need? Obvious sources of car bon are the earth's hydrocarbon resources and the earth's biom ass resources. Som e insights into the available hydrocarbon resources can be gained from figure 10.

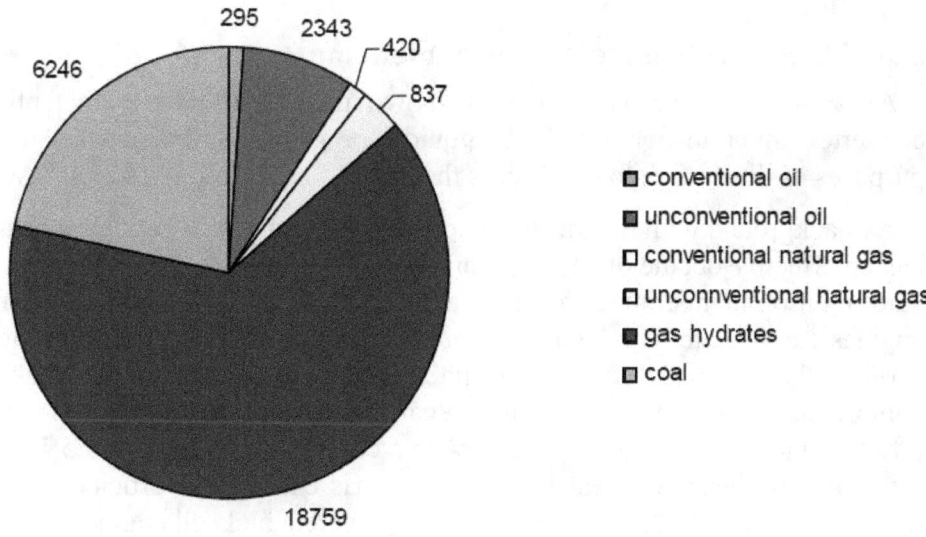

Source: H.H. Rogner "An Assessment of World Hydrocarbon Resources", Annu. Rev. Energy Environ. 1997 22: 217-262

Figure 10. Assessment of World Hydrocarbon Resources (Gtoe).

This figure summarizes the find ings from a recent assessm ent of world hyd rocarbon resources in units of Gtoe (10^9 tons of oil equivalent). It is clear that the current principle feedstock, conventional petroleum, represen ts a very sm all fraction of the earth's potentially available h ydrocarbon resources. If it were possible to access all the hydrocarbon resources indicated by figure 10, then, from a purely energy perspective, the energy needs of the earth could be met for thousands of years. There would, of course, be profound environmental considerations associated with such an undertaking. In reality, most of the hydrocarbo n resources indicate d by figure 10 will no t be econom ically or technologically accessible. Nevertheless, if only a fraction of them can be harvested in a cost-effective and env ironmentally acceptable fa shion, then the long-term prospects for satisfying the world's transportation energy needs are quite promising.

The unconventional oil indicated in figure 10 includes oil shale and tar sands. Unconventional natural gas includes coal be d methane, gas from fractured shale, and tight-formation gas. In this regard, it has been estimated that recent advances in the use of hydraulic fracture techniques for the extractio n of gas from shale m ay increase the potentially available U.S. natu ral gas supply by about one-third. [16] This results in a U.S. natural gas supply of about 100 years at current usage rates. Gas hydrates actually fall under the category of unconvention al natural gas but have been separated out because of the unique chemical configuration in which they occur and the estimated size of the v ast store of gas contained in this configuration. If one could economically and safely access the gas hydrates indicated in figure 10, then the methane derived would have the potential to provide natural gas for thousands of y ears. This could have a direct im pact on transportation fuels, because natural gas can be straightforwardly employed to synthesize alcohol and hydrocarbon fuels. We will discuss this next.

Liquid fuels via steam reforming of methane

All of the hydrocarbon resources shown in figure 10 have the potential for being converted into transportation fuels (liquid or gaseous). Perhaps the simplest case for purposes of illustration to consider is the indirect conversion of natural gas into methanol.

As background with regards to practicalities for this sub-section discussion: methanol has an effective octane of 119 compared to 87–93 for gasoline. This results in a higher auto ignition temperature for methanol. As a result, methanol internal combustion engines can operate with higher compression ratios than can gasoline engines. As noted earlier, the thermal efficiency of internal combustion engines increases as the compression ratio increases. In this regard, Brusstar et al.[17] have operated a methanol-powered internal combustion engine with a compression ratio of 19.5:1 and demonstrated a thermal efficiency of greater than 40%. This exceeds the efficiencies of gasoline and diesel engines and also that of methanol-powered fuel cell engines. As compared with gasoline and diesel power, methanol internal combustion, depending on the achieved engine energy efficiency, would require more fuel due to its specific energy being about half that of gasoline. Gallon for gallon, methanol combustion produces about half the carbon dioxide of gasoline combustion. Hence, at equal engine energy efficiencies, it would produce about the same amount of carbon dioxide, because about twice as much methanol would need to be combusted. If the methanol engine energy efficiency were double that of the gasoline engine that it replaced, it would produce about half the carbon dioxide of gasoline combustion. On the basis of this, one could argue that methanol combusted in a high compression ratio engine could be a viable transportation fuel.

We here describe the most common way to make methanol. The process begins with steam reforming of methane (SMR), where methane (CH_4) is the principal constituent of natural gas. SMR is shown in table 7 to be the cheapest way to make hydrogen on a large scale. In an elementary sense, SMR proceeds according to the reaction

$$CH_4 + H_2O \Leftrightarrow CO + 3H_2. \quad (4)$$

This reaction requires a catalyst (usually nickel), and the steam is usually in the 800–900 °C range.

If one desires to maximize the production of hydrogen, the water gas shift (WGS) reaction

$$CO + H_2O \Leftrightarrow CO_2 + H_2 \quad (5)$$

is also employed to yield the overall SMR reaction for the production of hydrogen

$$CH_4 + 2H_2O \rightarrow CO_2 + 4H_2. \quad (6)$$

At this point it is helpful to remember that mass is conserved in a chemical reaction. Thus, knowing the mass numbers of each element in a reaction allows a simple determination of the quantity of feedstock needed to produce a particular product. The mass number of hydrogen is 1, carbon is 12, and oxygen is 16. If one expresses mass in tons, then the SMR reaction says that 16 tons of methane combined with 36 tons of water (steam) produces 44 tons of carbon dioxide and 8 tons of hydrogen. Said another way, each ton of hydrogen desired will require 2 tons of m ethane and 4.5 tons of water and will produce 5.5 tons of carbon dioxide. So far we have neglected the fact that the steam must be heated to about $800^0 C$ in order for the reaction to proceed. T he energy to do this is taken from the m ethane supply and consumes about 1.5 tons of m ethane and produces an additional 4 ton s of carbon dioxide per ton of hydrogen. Thus, for each ton of hydrogen produced, the net m ethane consumption is about 3.5 tons, and the net carbon dioxide production is about 9. 5 tons. This sim ple arithmetic allows one to estim ate the SMR feedstock (m ethane and water) require d and the carbon dioxide produced for a given amount of hydrogen.

For simplicity, we introduce a simple "product" energy efficiency η_p where

$$\eta_p = \frac{E_p}{E_{FS} + E_S} \ . \qquad (7)$$

In this equation, E_p is the calorific content of the product produced, E_{FS} is th e calorific content of th e feedstock consumed and E_S is the calo rific value of any supplemental energy that was supplied to the process. This efficiency measures only the energy of the fuel product produced relative to the energy that was required to produce it. It does not give credit to th e value of byproducts of the pro cess, such as cogeneration of electricity from waste heat or the calorific or sales value of chemical byproducts. While the energy and m arket value of the byproducts are im portant, the interest in this paper is the transportation fuel p roduct of a synthesis process. We will us e this simple product energy efficiency as a metric for comparing the various process considered.

Hydrogen has a chem ical energy content of about 142 MJ/kg, and methane has a chemical energy of about 55.5 MJ/kg. Since it takes 3.5 kg of methane to produce 1 kg of hydrogen, it follows that $\eta_p = .73$ when SMR is used to produce hydrogen. In practice, the energy efficiency for SMR for hydrogen production is found to be between .6 and .7.

Instead of producing hydrogen from SMR, one can produce m ethanol by sending the products of eq. (4) to a m ethanol synthesis reactor, which typi cally operates at a temperature of about 250 °C and pressures of 50–100 atmospheres and employs a catalyst to favor the reaction

$$CO + 2H_2 \rightarrow CH_3 OH \ . \qquad (8)$$

The actual chem istries of SMR and m ethanol synthesis are so mewhat more complicated than that shown above. However, for the purpose of this discussion, the simplified chemistry is adequate. I n practice, the energy efficiency from methane to methanol is found to be about 60%.

The capital cost for a methanol plant that implements the above chemistry is dominated by the capital costs for the m ethane reforming and syngas condition ing steps. These account for about 80% of the plant costs.[18] One can, therefore, obtain a rough estim ate of the capital cost for a m ethanol capability by estimating the capital cost for producing the needed hydrogen using SMR technology and increasing that estimate by 20%. For example, to do an in-kind replacem ent of the 13.4×10^6 BPD of petroleum used for transportation would require about 250 billio n dollars o f capital in vestment for a methanol output with an energy equivalent to the petroleum used by today's U.S. transportation system. If one used a hydroge n production method other than SMR, the capital costs would depend on the scheme chosen.

Various attempts have been m ade to elim inate the exp ensive reforming step in methanol production by developing processes that convert m ethane directly into methanol. However, after decades of work, direct m ethane conversion is not yet competitive with conventional processes. This is largely due to the fact that m ethane is a fairly chemically inert com pound. Further, as with all car bon sources for fuels, the chemical properties of methanol need to be considered in evalua ting its suitability as a fuel. For ex ample: its lo w volatility can caus e starting prob lems at cold tem peratures, it can dissolve certain materials typ ically used as engine seals, it is c orrosive to some metals, it is toxic and it is hydrophilic. These properties would need to be considered in assessing methanol as viable transportation fuel . Details of m ethanol's physical and chemical properties and m aterial safety data sheets can be accessed via http://cetiner.tripod.com/Properties.htm.

The SMR process can also be e mployed to make liquid transporta tion fuels other than methanol. One such approach utilizes the methanol process disc ussed above. There are well-established techniques for converting methanol to gasoline. In p articular, in the 1970s a process now called MTG (m ethanol to gasoline) was developed by Mobil Oil. Some insights into that proce ss can be found in Meisel et al [19]. The details of the process are too complex to cover here. However, it is efficient at converting methanol to gasoline. The overall energy efficiency (including recy cled heat) of the MTG process is about 90%. Thus, the energy efficiency of the SMR through MTG process is about 54%. One of the problems that has b een identified with this approach is that the g asoline product contains about 40% arom atic hydrocarbons (e.g., tolulene, xylenes, and trimethylbenzene), putting it in c onflict with curren t environmental laws. I t is problematic whether this process should be pursued as an alternative to petroleum - derived gasoline.

Liquid fuels from coal

It was shown above in illus trative detail that methanol or gasoli ne can be m ade by using methane as a feedstock. In that cas e, the approach em ployed was to cause the methane to react with hot steam so as to p roduce a n ew gas consisting of carbon monoxide and hydrogen. This gas is referred to as "syngas." The syngas was sent to a reactor that was designed to cau se the carbon monoxide and hydrogen to react so as to produce methanol. Adding the MTG process step provides a pathway to gasoline. There are many other feedstocks that can be em ployed for liquid fuel synthesis. This section will discuss the use of coal as a feedstock for the production of liquid transportation fuels.

The total recoverable coal in the United States is estimated to be 280 b illion tons (EIA 2003). At the present rate of c onsumption, this resource will last about 280 years. In light of the size of these reserves, it is reasonable to consider what would be required to utilize this resource as a feedstock for liquid tr ansportation fuels. There are well-kn own processes that accomplish this. One approach that has been considered is to pyrolyze (i.e., heat in an inert atm osphere) the coal in order to drive off the vol atile matter (tar, light oils, hydrogen, etc.) that m ight be amenable to distillation techniques for the production of liquid fuels. This process has been studied for years. It has been found that, because the volatile matter in coal is typically 20–40 percent, pyro lysis results in a large fraction of solid char, which is principally carbon. As a result, pyrolysis, while it has m any uses, has not been heavily utilized for fuel synt hesis, because fuel synthesis wishes to maximize the conversion of carbon into a hydrocarbon or alcohol fuel.

A second scheme attempts direct liquefac tion through hydrogenation of coal. In some sense this scheme can be viewed as rep lacing the iner t atmosphere in pyrolysis with a hydrogen atmosphere that can react with car bon in coal to produce m ore distillable liquids and less char. Friedrich Bergius and his colleagues studied this approach during the period from about 1912 through WW I and found that they could obtain reaction products that were m ostly distillable liquids. The process, however, was slow and required gas pressures up to 700 atmospheres to obtain a large fractio n of distillable liquids. In 1931, Bergius receiv ed the Nobel Prize in chemistry for his work on chem ical reactions under high pressure (this rec ognition included his work on direct coal liquefaction). The Bergius process for coal liquefaction was scaled to comm ercial production levels and contribu ted to the Germ an war effort during WW II. This high-pressure, direct liquefac tion process probably has th e highest carbon conversion efficiency of the variou s coal liquefaction schemes. However, the very high pressure needed to obtain large fracti ons of distillable liquids has discouraged widespread industrial application of this approach.

A third scheme also developed in Germ any in the 1920s by Franz Fischer and Hans Tropsch has become the method generally used to synthesize liquid fuels from coal. This method is called indirect liquefaction and is si milar to SMR, in that it involves gasif ying the coal to create a gas (called syn gas) that is predom inantly CO and H_2. This gas is essentially the basis for producing hydrogen by coal gasification mentioned earlier. In the present case, a different synthesis route from that described above for SMR is followed. The syngas is purified and then used as th e feedstock for a process known as Fischer-Tropsch or FT synthesis. Th e FT p rocess involves a complex catalyzed chemistry that produces a variety of products, including para ffins, olefins, alcohols, and carbon dioxide. While coal is primarily carbon, it is not a pur e chemical compound but rather is a m ix of substances, usually includi ng carbon, hydrogen, oxygen, su lfur, and nitrogen. For example, for bituminous coal the average formula is $CH_{.8}O_{.1}N_{.02}S_{.02}$. A proper analysis of the FT process applied to coal would deal with all of its constituents. For simplicity, we will assume that coal is just carbon. The basic FT synthesis process proceeds according to the reactions

$$nCO + 2nH_2 \rightarrow C_nH_{2n} + nH_2O , \quad (9)$$

$$nCO + (2n+1)H_2 \rightarrow C_nH_{2n+2} + nH_2O . \quad (10)$$

The first reaction p roduces olefins, or alkenes, (C_nH_{2n}), while the second reaction produces paraffins, or alkanes, (C_nH_{2n+2}). As an example, when n = 8, the reader will recognize the paraffin product as octane. Th e FT process always produces a m ix of olefins and paraffins. However, the details of the m ix depend on the reactor conditions and the catalyst used. The FT reactors are generally operated in a low temperature range of 200–240 $^{\circ}C$ or a high temperature range of 300–350 $^{\circ}C$. Low-temperature operation favors high molecular weight waxes, while high-temperature operation favors low molecular weight olefins. The route to diesel fuel is through the low-temperature process while the route to gasoline is the high-temperature process. In both cases, product upgrading is required.

Since the FT process requires an input gas consisting of carbon monoxide and hydrogen, any process that provides such a gas is a potential candidate for creating the input gas. However, it can be seen from the governing chemical equations that maximum use of carbon in FT synthesis occurs when the input gas contains about two hydrogen molecules for each carbon monoxide molecule. Thus, a process like SMR, which creates a mixture of three hydrogen molecules for each carbon monoxide, is not, by itself, a good candidate for preparing the input gas for FT synthesis. Some modification of the SMR process, such as adding pure oxygen (autothermal reforming), would be desirable in order to use methane as the feedstock for FT synthesis. Such approaches have been considered.[20] These methane conversion schemes can be expected to have thermal efficiencies similar to the 60% efficiency that is characteristic of SMR.

Other methods of syngas preparation for FT synthesis are well developed. One such approach is the gasification of coal. Because of their general importance in the field of synthetic fuels, we have included a brief discussion of coal gasification coupled to the FT synthesis process in appendix B. A more detailed discussion can be found in Probstein and Hicks.[2] The simple analysis presented in appendix B suggests that one ton of coal will produce two barrels of oil equivalent product. This is similar to the industrial experience. From this one can estimate the process energy efficiency. The calorific energy content of one ton of coal is about 2.8×10^{10} Joules, while the energy content of a barrel of oil is about 6.1×10^9 Joules. Since most of the energy for this process comes from the coal feedstock, one finds that the process energy efficiency of coal gasification coupled to FT synthesis will be about 44% for the liquid fuel product. There are, of course, other energy credits that can be given For example, FT is a very exothermic process, and perhaps a 7% energy credit is available from cogeneration of electrical power. Other products, such as heavy waxes, have potential value. However, our interest here is the process energy efficiency to the desired transportation fuel. That efficiency is about 44%.

The combination of coal gasification and FT to produce liquid fuel is generally referred to as "coal to liquid" (CTL) conversion. The CTL process is well understood and could, in principle, be employed to provide for all U.S. transportation fuels. Such an undertaking would involve significant feedstock requirements and would also present significant carbon dioxide management issues.

The feedstock issue is easily quantified. The current rate of coal production in the United States is about 10^9 tons per year. Most of this is used to provide the United States electrical generating capacity. Since U.S. electrical consumption is unlikely to decline, any effort to provide transportation fuel from coal will require coal in addition to that which is currently mined. The United States consumes about 5×10^9 barrels of oil per year for transportation purposes. At a conversion rate of two barrels of oil per ton of coal, total oil independence at today's transportation fuel consumption would require that an additional 2.5×10^9 tons of coal be mined each year, resulting in annual co al consumption of 3.5×10^9 tons. Hence, if we were to provide for our current uses of coal and fill our current transportation-related petroleum needs by a convent ional CTL process then the current coal reserves of 280 years would be reduced to about 80 years. A 1.4% annual growth in consum ption would reduce that to about 50 years and a 5% growth would reduce it to about 30 years. While these estimates are only approximate, it is clear that the use of the conventional CTL pr ocess to resolve any significan t fraction of the nation's transportation fuel problem would have a pr ofound effect on the nati on's coal resources in a relatively short time. Nevertheless, it is c lear that CTL could, in p rinciple, provide for U.S. transportation fuel needs for an extended period. This would roughly double the CO_2 production associated with tran sportation, because the CTL process itself prod uces about as much CO_2 as does the combustion of the FT-produced liquid fuel. Since the CTL process is centralized, most of the CO_2 produced by it could be captured. The captured CO_2 could be converted to additional fuel by pr ocesses that will be discussed later. This would be expensive but would reduce the am ount of coal that n eeded to be g asified. . Also, if permitted, the captured CO_2 could be sequestered. In this regard, one should keep in mind that after about 80 years one would have sequestered about half the carbon in the U.S. coal supply in the form of gaseous CO_2.

It is logical to inquire whether m odifications to the conventional CTL process could reduce its coal consumption. A significant amount of the coal consumption calculated above is driven by an inadequa te supply of hydrogen in the f eedstock for the FT process; see appendix B (the Gasificati on and Fischer-Tropsch Process) . This situation could, in principle, be improved by inje cting into the output of th e coal g asifier an app ropriate amount of supplemental hydrogen. The precise amount of hydrogen required depends on the detailed kinetics of the gasifier. The example discussed in appendix B indicates that the hydrogen available from gasification designed to optimize the production of hydrogen and carbon monoxide (the feedstocks for the FT process) is about 50% of what is needed for the optim um syngas for the FT proce ss. As discussed in appendix B, som e improvement in hydrogen production is obtai ned by using the gas water shift (WGS) reaction. This, however, is wasteful of carbon and has already b een included in the analysis that leads to the estimate of two barrels of oil per to n of coal. If one elim inated the WGS step and instead injected a suppl emental amount of hydrogen into the syngas that was equal to that being produced by the gasifier, then the resulting syngas would be ideal for the FT synthesizer. This step would reduce the coal requirem ent by about 33% for the same FT product, resulting in about three barrels of oil equi valent per ton of coal gasified. This rate of coal conversion leads to a CTL co al requirement of about 1.7×10^9 tons per year and a total coal requirem ent of about 2.7×10^9 tons per year. At this rate, the lifespan of the current U.S. coal supply woul d be reduced to about 105 years, assuming no additional growth in use. This is 25 years longer than CTL without supplemental hydrogen. The price one pays for this coal supply lifetime extension is the need to provide about 5×10^5 TPD of supplem ental hydrogen. This is about the am ount of hydrogen that would be needed to provide for a hydrogen internal combustion engine solution to the entire transportation fuel problem. The supplem ental hydrogen infrastructure would itself be a massive complex.

It is straightforward to estimate the capital cost of the combined CTL and supplemental hydrogen infrastructure. Using Na tional Mining Association es timates, one finds that a

13.4×10^6 barrel per day conventional CTL capacity would cost about 900 billion dollars to construct.[21] The provision of supplemental hydrogen should reduce the capital cost of the coal gasification component of that capability by about 300 billion dollars. The cost of hydrogen production for several different technologies is given in table 7. If one scales these hydrogen production capital costs to those associated with a 500,000 TPD capability, one can estimate the capital cost associated with CTL involving supplemental hydrogen. These estimates are given in table 8.

Hydrogen production technology	ROM of capital costs for 500,000 TPD hydrogen production capability (billion dollars)	ROM of capital cost of 13.4×10^6 barrel per day CTL capability with supplemental hydrogen (billion dollars)
Steam Reforming of Methane (SMR)	157	757
Biomass gasification (BG)	450	1050
Coal Gasification (CG)	467	1067
Nuclear reactor Sulfur – Iodide Cycle (SI-MHR)	967	1567
Conventional Electrolysis Powered by Nuclear Reactor (CE-NR)	1800	2400

Table 8. Estimates of capital costs associated with providing supplemental hydrogen to a CTL capability and the capital costs of the resulting CTL capability when the CTL capacity is 13.4×10^6 barrels per day.

The third column in table 8 should be compared with the conventional CTL capital cost estimate of 900 billion dollars. Of the supplemental hydrogen production schemes considered, only the SMR scheme indicates a net reduction in CTL capital cost. The SMR supplemental hydrogen production would increase CO_2 production by about 5×10^6 TPD and methane consumption by about 1.3×10^6 TPD. This would roughly double the CO_2 produced by the transportation sector and double the rate of consumption of natural gas. The use of coal gasification to produce the supplemental hydrogen makes no sense, because it would consume coal faster than CTL without supplemental hydrogen. The two nuclear schemes considered would increase the lifetime of the coal supply by perhaps 25%, but would require a much larger capital investment than using an SMR approach to supplemental hydrogen or just using a conventional CTL approach without supplemental hydrogen. The gasification of biomass to produce the supplemental hydrogen would extend the lifetime of the coal supply by perhaps 25%, but would have higher capital costs than using SMR or just conventional CTL. It would also introduce the complexity of having to run two separate large gasification infrastructures.

From the above discussion it would seem that the injection of supplemental hydrogen into the CTL process, while possible, would not be practical or cost effective. Another option to consider would be to perform th e entire liquid fuel production using carbon neutral biomass gasification. We will consider that option next.

Liquid Fuel via Biomass Gasification

In the above two sub-sections we have cons idered methane and coal as sources of the syngas used for liquid fuel synthesis. These approaches were shown to be capable of providing for the full U.S. transportation fuel needs for significant periods of ti me. Such use would greatly increase the consumption of coal and natural gas, thereby reducing the lifetimes of those reserves. They would also have the side eff ect of producing large amounts of additional carbon dioxide.

The synthesis of liquid fuels using the FT process requires a feedstock consisting of carbon monoxide and hydrogen. The preferre d ratio of hydrogen m olecules to carbon monoxide molecules in the feedstock is 2:1. The FT reactor does not care how the syngas is produced as long as it has been prepared so as to be free of impurities that might poison the FT catalysts and enters the reactor with the proper temperature, pressure and mixture. Therefore, biomass (e.g., hardwoods, softwoods, grasses, crop residues, etc.) gasification provides another potential option for the source of the FT fe edstock. The use of biom ass feedstocks, if done in a sustainable fash ion, avoids much of the carbon dioxide management issue, because the carbon dioxide that is released during the production and use of these fuels com es from the atm osphere. Because of this, the biomass feedstock approach is said to be carbon neutral. This stat ement is true to the extent that fossil f uels are not use d to grow and harvest the biom ass. In this sub -section we will discu ss the extent to w hich biomass gasification can a ddress the transportation fuels problem . The combination of biomass gasification with FT synthesis is referred to as BTL.

The chemical makeup of biom ass has been well studied and catalogued (see for example http://www1.eere.energy.gov/biomass/feedstock_databases.html) Dry biomass is composed of cellulose, hem icellulose, lignin, protein, and ash. The fractions of these compounds vary with the type of biom ass. However, on average, the co mposition is as follows:

Compound	Weight Percent
Cellulose (45% carbon)	44
Hemicellulose (48% carbon)	27
Lignin (40% carbon)	20
Other	9
Total	100

Table 9. Average chemical composition of dry biomass (hardwoods, softwoods, grasses, crop residue, etc.).

39

From this table we see that, on averag e, carbon accounts for about 41% by m ass of biomass. We can use the previous analys is of the CTL process to gain som e understanding of the a bility of biom ass gasification to contribute to a solution of the transportation fuels problem. The most plentif ul form of coal in the United States is bituminous coal, which has an average carbon content of about 70%. Thus, one can expect that the production of FT liquids fr om the gasification of biomass will req uire about 1.7 times the mass of coal needed for the equivalent liquid product production from the CTL approach. This results in a dry biomass requirement of about 4×10^9 tons per year to produce a FT product with energy content approxim ating that currently consum ed in U.S. transportation fuel. A recent study by the Department of Energy and the Department of Agriculture concluded that about 1.4×10^9 tons per year of biom ass could be recovered from U.S. forest resources and agricultural resources in a sustai nable fashion without adversely impacting the forest and agriculture industries. [22] This would be a significant undertaking and involve a "more than seven-fold increase in production from the amount of biomass currently consumed for bioenergy and biobased products." One can conclude from this that an aggressive biom ass gasification program could produce at m ost about one-third of the current U.S. transportati on fuel requirement. W hile this is a crude estimate, it suggests that it will be difficult for biomass gasification alone to provide, on a sustainable basis, the liquid fuel equivalent of the 13.4×10^6 BPD of oil now used for the current U.S. transportation system.

It should be noted that biom ass gasification chemistry is somewhat different from coal gasification chemistry. This is because cellulose is the basic building block of biom ass, resulting in a chemical structure based on the $C_6H_{10}O_5$ complex. This is quite different from the average chem ical formula for bituminous coal, CH_8O_1. As a result, the output from the biom ass gasifier includes significant wa ter, methane, and tar. It is, ther efore, necessary to include a reform ing step that converts the m ethane and tar into carbon monoxide and hydrogen prior to entering the gas cleanup apparatus and the FT reactor. An example of how to perform this gasifi cation and reforming has been reported by Spath et al. [11] and is the basis for the capital costs quoted in table 6 for hydrogen production by biomass gasification. In that case, the gas that leaves the tar reformer has a hydrogen to carbon m onoxide ratio of 2.03:1, which is about what is required for complete consumption of carbon monoxide in an FT reactor. Unlike CTL, where there is substantial experience regarding capital costs, there are no large commercial BTL plants from which one can scale BTL capital costs. However, it is known that about 70–75% of the capital costs in the production of syngas from gasification are associated with the syngas production, gas conditioning, and power generation, while 10–20% are associated with FT synthesis and product upgrading. If one assum es that a similar breakout of costs will occur for BTL systems, then one can u tilize the analysis reported by Spath et al. [11] to provide a rough estim ate for the capital cost for a BTL f acility that produces 13.4×10^6 barrel per day of FT liquid product. We saw above that this will require about 4×10^9 TPY of dry biomass, or 1.1×10^7 TPD. The Spath et al. [11] estimates for hydrogen production were based on a 2000 TPD biom ass input to the ga sifier. If we simply scale the Spath et al. [11] capital cost estimate of 154 m illion dollars to the required biom ass input, we obtain 893 billion dollars, which is quite similar to the CTL estimate of 900 billion dollars for the same quantity of FT liquid product. In reality, a BTL capital inves tment would likely be limited to about 300 billion dollars because of the limit on the availability of biomass that was mentioned above. As noted earlier, this capability would provide for about one-third of the needed fuel.

It is straightforward to make a rough estimate of the BTL process energy efficiency. It takes about 1.7 tons of biomass to produce two barrels of oil equivalent liquid by the BTL process. A ton of dry biomass has a typical calorific value of about 1.5×10^{10} Joules, while a barrel of oil has a calorific value of about 6.1×10^{9} Joules. These numbers suggest that the BTL process has an energy efficiency of about 47%. As expected, this is comparable to the process efficiency of 44% found for the CTL process.

There are several other methods for producing transportation fuels from biomass. A method employing fermentation technology will be discussed in section 7.

7. Enzymatic Approach to Producing Fuels from Biomass

It was pointed out previously that the car bon in biom ass is cont ained principally in carbohydrates and in lignin. Th e carbohydrates (cellulose, he micellulose, and starch) are composed of sugars or polym ers of sugars . Lignin is not a car bohydrate and is not amenable to the processes describe below. It has been known for m illennia that certain sugars can be fermented (anaerobically decomposed into alcohols and carbon dioxide) in the presence of enzymes (complex proteins that are produced by cells and act as catalysts in specific biochemical reactions). The reduction of carbohydrates to these sugars is a key step in the production of alcohol fuels. Perhap s the best known exa mple of such a fuel is ethanol synthesized from cornstarch. The basic process here i nvolves the hydrolysis (decomposition by reaction with water) of cornstarch $(-C_6H_{10}O_5-)$ and the fermentation of the resulting Maltose ($C_{12}H_{22}O_{11}$) to produce ethanol (C_2H_5OH). The overall reaction can be written as

$$2(-C_6H_{10}O_5-)+2H_2O \rightarrow 4C_2H_5OH + 4CO_2 . \qquad (11)$$

It follows from eq. (11) that 57% o f the co rn starch is converted into ethanol. Since about 57% (by weight) of a bushel of corn kernels is starch, it follows that about 33% can be converted into ethanol. A typical bushel of corn kernels weighs about 56 lbs. Thus, a bushel of corn yields about 18.5 lbs, or 2.8 gal. of ethanol. The co rn to ethanol process requires about 64% of the energy contained in the ethanol product. [2] This energy is typically provided by burning natural gas. Th e energy content of bone-dry corn is about 14 MJ/kg. The energy content of methanol is about 30 MJ/kg. Thus , the sim ple energy efficiency of the corn to ethanol process is about 46%. The capital co st for a large corn ethanol plant is estimated to be about $1 per gallon per year.

The ethanol equivalent of U.S. oil cons umption for transpor tation is about 3×10^{11} gallons per year. If corn -based ethanol were to provide this, it would require about 10^{11} bushels of corn. The typical productivity of an acre of corn is about 130 bushels. Thus, the process described above would require about 10^9 acres of corn production to yield the ethanol energy equivalent of today's transportation fuel needs. This represents about forty percent of the landmass of the United States and exceeds the estimated arable land (4×10^8 acres) available in the United St ates. Current annual corn production in the United States is about 10^{10} bushels. Most of that is used to f eed livestock, poultry, fish, and people. These results indicate that enzymatic production of ethanol using only corn starch as a feedstock is not viable as an approach to pr ovide for U.S. transportation fuel needs. If ethanol were to be the fuel of choice, then other feedstocks must be considered.

Corn starch is a sm all fraction of U.S. biom ass. It can be seen from table 8 that abo ut 71% by weight of biom ass (plants, crops, trees, etc.) resides in cellulose and hemicellulose, both of which are polym ers of sugars. If these polym ers can be decomposed into sugars that are ferm entable, then the basic process discussed above for the production of ethanol can be utilized. Th is opens up a broad array of crops beyond corn. Obvious candidates are grain crops, whos e seeds, like corn, are high in starch and are easily hydrolyzed to fermentable sugars. Another obvious candidate is sugar crops, because the sucrose from these crops is easily hydrolyzed to fermentable sugars. After the seeds and sucrose have been rem oved from grain crops and sugar crops, considerable carbon remains in crop residue in the for m of cellulose, hemicellulose, and lignin. There are also the lignocellulosic crops, such as grasses, shr ubs, and trees. Th e lignocellulosic residues from crops and lignocellulosic crops have proved to be difficult as feedstocks for fermentation. However, recent developments in the areas of cellu lases and xylases show promise for hydrolyzing the cellulose and hem icellulose found in biom ass. Considerable work remains to b e done to sho w that the required c ellulases can be produced economically on a scale required for synthetic fuel production. However, if one assume s that the ongoing work will b e successful, then the application to lignocellulosic biomass will follow a total reaction path similar to that shown in equation (11), but acting on the polysaccharides that make up cellulose and hemicellulose. As a result, 1 g of cellulose or one gram of he micellulose will yield abou t 0.57 g of ethanol. Since cellulose plus hemicellulose constitute about 71% of the biomass, it follows that a ton of biomass could, in principle, produce about 0 .4 tons of ethanol.

The process of cellulosic ethanol production is much more energy intensive than that for corn ethanol production, and the production plants will be much more expensive to build than are the corn ethanol plan ts. A plant design stud y was performed recently to compare the costs of building a ligno cellulosic ethanol plant to for the costs of building a corn ethanol plant.[23] The lignocellulose plan t utilized a fluidized bed combustor fed by the lignocellulose feedstock to provide for the process energy requirem ents. The corn ethanol plant employed the much less expensive natural gas boiler to provide the process energy. A 50 million gallon per year lignocellulose plant was estimated to cost about 194 million dollars. The sa me capacity corn ethan ol plant was estimated to cost 48 m illion dollars. As noted ear lier, the corn ethanol app roach, while much less expensive, is not viable as a total so lution, because there is insu fficient landmass to gro w the neces sary corn. The lignocellulose plant was predicted to produce 79 gallons of ethanol per dry ton of lignocellulose feedstock. If one takes the biomass feedstock as having an energy content of 17 MJ/kg and ethano l as having an energy value of 80 MJ/gal one obtains a simple energy efficiency of about 41% for the lignocellulose plant. This is comparable to the simple energy efficiency calculated above for a typical corn ethanol plant.

The information given above allows one to estimate the num ber of biom ass acres needed to provide from lignocellulose the et hanol energy equivale nt of today's U.S. transportation fuel as a function of the li gnocellulose biomass yield per acre. This estimate is plotted in figure 11.

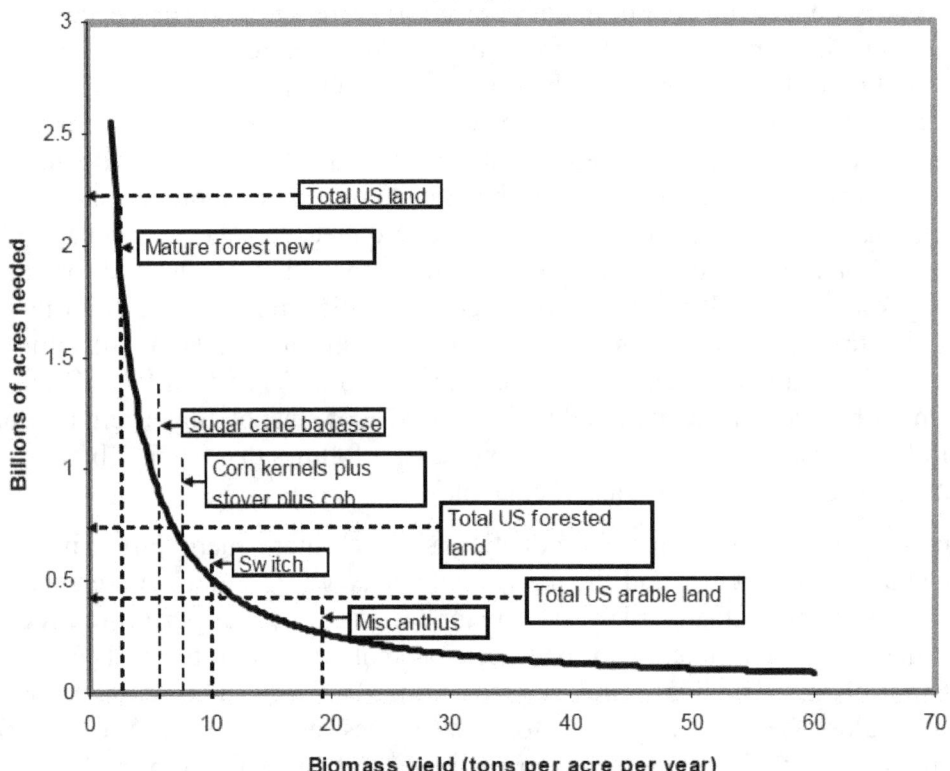

Figure 11. Biomass acres needed to provide the ethanol energy equivalent of today's US transportation fuel from ligno-cellulose vs the ligno-cellulose biomass yield per acre.

The horizontal dashed lines in figure 11 indicate the total U.S. landmass, the total U.S. forested landmass, and total U.S. arable landm ass. The vertical dashed lines ind icate the annual new wood growth per acre added to a ma ture forest, the sugarcane bagasse yield per acre, the corn kernel plus stover (leaves and stalks) plus cob yield per acre, the switch grass yield per acre, an d the miscanthus yield per acre. Th e intersection between these vertical lines and the solid curve provides an estimate of the number of acres needed for a particular feedstock. One can see from figure 11 that there is not enough new growth in mature forested lands to provide the wood to obtain sufficient ethanol from the U.S. forests. The acreag e required for s ugar cane bagasse or for corn, including kernels, stover, and cob, exceed s the total U.S. arab le landmass. The perenn ial grasses, sw itch grass, and m iscanthus require approximately the total arable landm ass. These perennial grasses can grow on land that today would be considered marginal, thereby reducing the impact on food production. However, the biom ass yield will likely decline on m arginal land. It see ms clear th at a single crop to etha nol solution will have great dif ficulty in solving the full transportation fuel problem while at the sam e time providing for the critical traditional products from agricultural and forested lands.

It was noted earlier that a DOE/USDA study concluded that, with som e effort, about 1.4 billion tons of biomass per year could be obtained from the forest and agricu lture industries without adversely im pacting the ab ility of those industries to m eet their

44

traditional and essential responsibilities.[22] If cellulosic ethanol becomes a reality, then 1.4 billion tons of biom ass could produce about 360 million tons, or about 10^{11} gallons, per year of ethanol. This is about one-third of the ethanol requ ired for a full solution to the transportation fuels problem. It is estim ated that the capital cost of a larg e cellulosic ethanol plant would be about 4 dollars per gallon per year. [23] This leads to a capital cost for converting 1.4 billion tons of biom ass into cellulosic ethanol of about 400 billion dollars. This cost should be com pared to the estimated capital cost of 300 billion dollars for gasifying the sam e biomass and producing FT liquids with roughly the sam e energy content. Thus, within these rough estimates, the biom ass to ethanol approach appears have capital costs that are about 30% hi gher than the BT L approach. The FT product from the BTL approach would likely be m ore compatible with the current gasoline and diesel infrastructures and with the FT f uels that would need to be produced from other feedstocks that would be needed to m ake up the two-thirds of the fuel that biom ass could not produce in a sustainable fashion.

A factor that we have not discussed involves the non-solar energy that must be invested to grow the biomass. If that energy must come from the biomass-produced fuel, then the energy content of the biom ass fuel must be many multiples of the biom ass-produced energy invested to grow the bi omass in the f irst place, if the proce ss is to be self sustaining. Also, if the non-so lar energy needed to grow, pr epare, and distribute the biomass comes from fossil fuels, then the ove rall process is not carbon neutral. T hese topics are beyond the scope of this paper but have been m atters of some controversy in the bio-energy field for some time.

A DOD-only solution using biom ass requires only one 50th of the acreages shown in figure 11. In principle, a DOD solution could be accomm odated with any of the biomass sources indicated in f igure 11. However, the requir ed landmasses are still larg e. For example, a corn kernel solution for DOD would require about 30 million acres. The entire U.S. corn production involves about 90 million acres, of which only about 4 million acres are currently dedicated to ethanol production. Even a DOD -only solution would have a large impact on the U.S. forestry and agriculture industries.

8. Fuels from Biomass Oils

In the above discussion we have discu   ssed the topic of synthesizing fuels from biomass. There are also natural oils that occur in biom ass (seeds, animal fats, etc.) that provide another potential approa ch to alternative transpo rtation fuels. Fuels m ade from natural oils are genera lly referred to as biodiesel fuel s. About 95% of biomass oils are triglycerides (three fatty acids plus glycerol). There exists a mature biorefinery industry that produces a large number of products (soa ps, detergents, lubricants, solvents, etc.). The basic technology em ployed here is well   developed. The proc ess consists of a crushing step where the bio oil is separated from the feedstock (e.g., soy beans) and an oil conversion step that converts the triglycerides in the bio oil into bi odiesel (methyl ester) and glycerine. In the oil conversion proce   ss, 1 kg of bio oil produces about .97 kg of biodiesel and .2 kg of glycerine and cons   umes about .1 gram   of m ethanol. The predominant bio oil f eedstock used in the United States is   soybean. W e will use soy beans as an illustrative example.

Bio oil con stitutes about 19% by weight of the soybean. Thus, a 1 kg bio oil input requires about 5.3 kg of soybean. The calorific energy content of soybean is about 18.7 MJ/kg. The calorific energy of biodiesel is   about 37 MJ/kg, and the calorific energy content of methanol is about 19.7 MJ/kg. A detailed study of the life cycle of biodiesel produced from soybean found that about 17%   of the biodiesel product energy was required to be added in the crushing and conversi on steps of the process. [24] From these numbers one finds that the simple process-energy efficiency of the biodiesel process from soybean feedstock to biodiesel product is about 35%.

The biodiesel lifecycle report m entioned above provides som e data regarding the capital cost of constructing biodi esel plants. While there is considerable variability in the data, a capital cost of about $1 per gallon per year of biodie sel product is representative of the cited capital costs. Thus, if one were able to provide the bio oil feedstock needed to produce the $2x10^{11}$ gallons per year needed for a biodiesel solution to the national transportation fuels problem, the capital cost involved would be in the vicinity of $2x10^{11}$ This is comparable to the capital cost estimated earlier for methanol solution using SMR.

There are o ther plant and vegetable oils that can be used for feedstock for biodiesel production. Among these are corn oil, sunflower seed oil, peanut o il and palm oil. The biodiesel yields for these crops are well known and are listed in table 10.

Oil Crop	Biodiesel yield in gals/acre	Annual MJ/acre
Corn	14	1,746
Soybean	40	4,988
Peanut	90	11,223
Avocado	225	28,0575
Palm	500	62,350

Table 10. Biodiesel yield from several bio oil crops.

One could use table 10 to estim ate the number of acres that would be needed by each of the various oil crops in or der to satisfy the transportation fuel requirements. It is simpler to just com pare the o il crop yields with the ligno cellulose crop ethanol yields discussed above. These yields are summarized in table 11 for several lignocellulose crops.

Lignocellulose Crop	Ethanol yield in gals/acre	Annual MJ/acre
sugar cane bagasse	475	37,968
lignocellulose corn	634	50,625
switch grass	792	62,280
miscanthus	1,426	114,000

Table 11. Ethanol yield for several lignocellulose crops.

The lignocellulose energy yields are seen to be com parable to or greater than the biodiesel crop yields. Therefore, from the point of view of solving the national transportation fuels problem , biodiesel a pproaches will face even m ore serious difficulties than those faced by etha nol approaches. This is further complicated b y the fact that the high-yield biodi esel crops such as avocado and palm are very lim ited regarding where they can be grown. The NREL biomass oil analysis study m entioned above concluded that the ulti mate supply potential for bio o il fuel is about 10 billion gallons per year. This represents about 5% of the U.S. transportation fuel need. The addition of biofuel from animal fats does not significantly change this result. Thus, while bio oils and biodiesel fuel m ay have important niche applications, they are not a solution to the national transportation fuel problem.

There is one potential, although unlikely, exce ption to the bio oil conclusion. It has been conjectured that certain algae may produce bio oil yields greater than 10,000 gallons per acre. This projected yield derives from results obtained in the Departm ent of Energy Aquatics Species Program (ASP), where algae strains we re found having natural oil content as high as 60%. [25] The ASP was in itially motivated by the idea of using a lgae ponds as a way to sequester CO_2 emissions from smokestacks. It was envisioned that CO_2 would be separated from the stack gase s and bubbled through a flowing algae pond, where it would be taken up by the algae. The experimental data indicated that certain algae strains were very good, unde r the proper circum stances, at converting CO_2 and sunlight into oil. As a result, the ASP changed its thrust from that of CO_2 sequestration to biodiesel production. If the projected yield were actual ly realizable on a large, sustainable, and econom ically viable scale, it would have a significant im pact on the viability of biomass-derived fuels as a solutio n to the na tional transportation fuels problem. If successful, the landm ass required to meet the nationa l transportation fuel need would be about one percent of the U.S. landmass.

The ASP project found that the con ditions that promote high productivity and rap id algae growth and the conditions th at induce high oil accumulation seem to be mutually exclusive. Therefore, the conjectured high yields were never achieved in the ASP. There is, however, som e hope that genetic m anipulation of prom ising algae strains m ight produce strains that are simultaneously capable of both high growth and high lipid (oil) synthesis. The new strains would also have to be able to maintain these properties in the demanding operational environments in which they would ultimately function. Research to achieve these goals is high ri sk, but, if it were successfu l, it could have a very large impact. The data needed to assess the efficiency and expected capital cost associated with this approach remains to be obtained.

9. Synthetic Fuel from Atmospheric Carbon Dioxide

Among the factors motivating interest in alternative transportation fuels is concern over the buildup of CO_2 in the atmosphere. In this regard, an obvious approach to consider is the possibility of removing the CO_2 from the atmosphere at the rate that transportation fuels are adding CO_2 to the atmosphere and converting that CO_2 into transportation fuel. This is similar to the biomass approach discussed in the previous sections, except that, in this case, the removal scheme for CO_2 involves artificial filtration rather than photosynthesis. The extracted CO_2 could be converted into a transportation fuel which, when consumed, would return the CO_2 to the atmosphere. The cycle would, therefore, be CO_2 neutral, provided that the energy source needed to extract the CO_2 and to make the synthetic fuel was not itself a CO_2 producer. One approach for doing this was proposed by Stucki et al. [26] This approach involves an absorption/desorption process in which a K_2CO_3 (potassium carbonate) solution formed through CO_2 absorption by KOH (potassium hydroxide) is fed to an electrolyzer where CO_2 and KOH are regenerated by the overall reaction

$$4H_2O + 2K_2CO_3 = 2H_2 + O_2 + 4KOH + 2CO_2 . \quad (12)$$

The regenerated KOH is returned to absorb more CO_2, and the regenerated CO_2 is converted into methanol through the mildly exothermic reaction

$$CO_2 + 3H_2 = CH_3OH + H_2O . \quad (13)$$

Using experimental results, Stucki et al. designed a one-square-meter module consisting of an array of hollow fibers (240 mm inner diameter) through which a 0.5 molar KOH solution would flow at 2 cm per second. The array of fibers provided an absorption membrane area of 18 square meters with a total fiber length of 23,000 m. The air was assumed to flow through the array at 3 m per second. The experimental data indicated that this array would remove 3.7 kg of CO_2 per hour from an atmosphere containing 350 ppm of CO_2. The air was predicted to undergo a 50% CO_2 depletion as it passed through the array. An electrolysis cell would be used to regenerate the KOH and CO_2 according to reaction (12), and in the process would also generate one-third of the hydrogen needed for the methanol synthesis reaction (13). Additional hydrogen will be needed because, for maximum carbon consumption, reaction (13) requires three hydrogen molecules per CO_2 molecule, while equation (12) generates only one hydrogen molecule per CO_2 molecule. The electrolyzer that would regenerate the, CO_2, KOH and also supply the needed hydrogen would require about 30 kilowatts of electrical power for the case studied by Stucki et al.

A variation on this approach has recently been studied by Los Alamos National Laboratory (LANL). The LANL scheme, designated as "Green Freedom," employs a K_2CO_3 solution to absorb CO_2 to form potassium bicarbonate ($KHCO_3$).[27] Absorption of greater than 95% is claimed. The bicarbonate solution is then sent to an electrolyzer, where CO_2 is regenerated along with one molecule of H_2 for each molecule of CO_2. The CO_2 is then reacted with hydrogen to synthesize methanol according to equation (13). As with the Stucki scheme, supplemental hydrogen must be added to achieve the stoichiometric coefficient required by eq. (13). The methanol produced is then converted to gasoline by the Mobil MTG process mentioned earlier.[19]

Since the Green Freedom results are recent, we will use them to estimate the energy efficiency of the synthesis of transportation fuel from atmospheric CO_2. The Green Freedom program proposes to employ conventional electrolysis powered by nuclear reactors. Since the individual processes involved in Green Freedom are well understood, it is possible to make reasonably accurate estimates of the process efficiency and of the capital costs involved in building the production plants. The CO_2 feedstock has no calorific value. However, it is found that the nuclear fuel must provide about 5.7 Joules of thermal energy for each Joule's worth of gasoline produced. This leads to a simple process-energy efficiency of about 18%. This efficiency could be raised to perhaps 25% if more energy-efficient methods (e.g., SI thermal cycle or high-temperature electrolysis) for producing the supplemental hydrogen proved to be viable. The Green Freedom program estimated the nuclear reactor capital costs based on using the expected costs of the Westinghouse AP1000 third generation pressurized water reactor as the process-energy source. The other components are commercially available, and their capital costs are known. The total system capital cost was found to be about 290,000 dollars per barrel per day of oil equivalent product. This leads to an estimated capital cost of about 4 trillion dollars for a capability equivalent to the 13.4×10^6 BPD currently used for transportation. About 58% of the capital cost is due to the nuclear reactor costs. The scheme would require about 1600 three-gigawatt thermal nuclear reactors. For reference, there are about 444 nuclear power reactors in operation around the world.

Other proposals for converting CO_2 into transportation fuels have been considered. For example, the CO_2 emitted from fossil-fired smokestacks has much higher density than atmospheric CO_2, thereby simplifying CO_2 extraction. While such an approach is not CO_2 neutral, it could substantially reduce the net CO_2 introduced into the atmosphere. Another approach that has been suggested is to extract carbon from the oceans, where the carbon content is much higher than in the atmosphere. This has been suggested as a potential approach for fuel synthesis at sea to support deployed naval forces. A brief discussion of this will be found in Coffey et al.[7] For these various approaches, as with the Stucki et al. and Green Freedom approaches discussed above, the regeneration of the CO_2 upon extraction and the provision of the needed supplemental hydrogen for fuel synthesis will be energy intensive and have capital costs similar to those given above, because the same amount of CO_2 must be regenerated, and the same amount of supplemental hydrogen must be provided.

10. Exploitation of Oil Shale

Production of liquids from oil shale appear s on m ost lists of potential alternative transportation fuels. This is becau se there is a lot of oil shale. The m ost promising U.S. deposits are found in W yoming, Utah, and Colorado. The m ost recent extensive assessment of the prospects for shale oil was published by the Office of Technology Assessment in 1980.[28] This assessment is still quite relevant.

While oil shale's organic cont ent is relatively low (about 20 % that of coal), the W orld Energy Council (WEC) puts U.S. proved recovera ble and estimated recoverable oil from shale to be about 6.5×10^{11} barrels.[29] If this o il were reco vered and u sed to m eet the current annual U.S. petroleum requirement for transportation fuels, then the supply would last about 140 years. A 1.4% annual grow th in dem and would reduce that num ber to about 90 years, and a 5% annual growth in de mand would reduce the time to depletion to about 50 years. These num bers indicate that oil shale is a significant potential source of transportation fuel.

The approach to fuels from oil sh ale is qualitatively different from the approaches discussed in the previous sections. Most of the organic content of oil shale is found in a substance called kerogen. Kerogen is not a m ember of the petroleum family and is not soluble in conventional petroleum solvents. As a result, it is not recoverable by solvent extraction. The standard practice for recoverin g oil from oil shale is to m ine the shale, move it to a retort, and pyrolyze it. This practi ce is referred to as "surface retorting." The oil yield from this process has substantial va riability, ranging from 10 gallons per ton to 60 (or m ore) gallons per ton of m ined shale. Yields greater than 25 gallons per ton are believed to be required for econom ically viable production of shale oil.[25] If one assumes a yield of 30 gallons per ton, then it would be necessary to mine about 2.3×10^9 tons of oil shale per year to m eet current U.S. petroleum needs. This is about the sam e as the estimate given earlier regarding the am ount of coal that would need to be m ined to meet U.S. petroleum needs through the CTL process.

Since oil shale is predominantly inorganic, most of the m ined material would have to be disposed of. The large volumes of shale needed to be mined and the need to dispose of a large was te stream have been major im pediments to commercial exploitation of oil shale. This has led to s everal efforts to examine *in situ* retorting of the shale, where th e shale would be pyroly zed in place, thereby reducing the m ining and waste disp osal requirements. The basic idea is to drive out and condense the volat ile components of the kerogen and extract the oils and gases by c onventional petroleum retrieval schemes. One approach to this is to fracture the shale and com bust the upper layer of the kerogen-containing shale, thereby for ming a reto rting and vaporization zone below the combustion zone. The vapor produced is conde nsed, and the resultant oil, water, and gases are pumped out. This proces s is m ore energy intens ive than the surface retorting process but has the advantage of greatly re ducing the waste disposal p roblem. It also raises considerable concerns regarding cont amination of ground water. A variant of this scheme has been studied by Shell Oil co rporation through its Mahogany Research Project.[30] In this s cheme, electric heaters are ins erted into the shale and gradually heat a volume of shale to about 700 °F. In order to keep the oil produced within a fixed volume, the ground surrounding this volume is frozen. Tests were recently completed (2004) by applying this scheme to a sm all volume (about 1200 square feet of surface area). About 1700 barrels of light oils were recovered al ong with associated gases. Approximately one-third of the energy released was in th e gases produced and two-thirds in the oil produced. This approach appears attractive, because it would seem to have a m uch lower environmental impact than the conv entional techniques for recovering shale oil. Shell asserts that the process is less expensive th an the conventional approaches for extracting shale oil.

It is inf ormative to es timate the ca pital investment that would need to be m ade to recover the shale oil. In 1980 it was estim ated (OTA) that a 50,000 barrel per day oil shale facility using underground mining and surface retorting would involve a capital cost of about 1.7 billion dollars (1980 dollars). If one simply scales that cost by inflation, one arrives at an estimate of about one trilli on dollars (2005 dollars) for a 13.4 m illion barrel per day capability from surface retorting. This compares with an estimate of 900 billion dollar for p roducing the same quantity of oi l equivalent liquid us ing the CTL or BTL approaches.

To the author's knowledge, there is no estim ate available for the *in situ* approaches. However, one can g ain some insight in to the capital costs associ ated with th e Shell approach. If the Shell approach to *in situ* retorting utilizes electric heating, then substantial capital costs will be associated with meeting the electric power demands. One can gain some insight into thos e capital costs by examining the power dem ands. Shell asserts that the energy return on energy invested (EROEI) for their process is about three. Most of the energy invested (EI) is associated with the electrical power needed to heat the shale. The energy return ed (ER) rep resents the energy in th e extracted oil and gas. A barrel of oil contains about $6x10^9$ Joules, and the United States consumes about $13.4x10^6$ barrels of oil per day for transportation purpos es. If ER = 3E I, then these numbers imply that, in steady state, the energy inve sted per second would be about $3.4x10^{11}$ Watts. This is comparable to the average U.S. el ectric power consumption of about $5x10^{11}$ Watts. Considering the large amount of power needed and the rem ote locations of oil shale reserves, the Shell app roach would likely require the construction of new generating capacity. If we assum e that th e power is supplied by power plants at a cap ital cost of $1,500/KW_e$, then the capital cost for providing the electrical generating capacity needed to heat the oil shale at the rate required to supply the current U.S. transportation dem and would be about 500 billion do llars. This capital cost do es not include the drilling and refrigeration capital cos ts. Thus, the 500 billion dollar co st estimate represents a lo wer bound on the actual capital cost. The further ad ditional capital costs associated with oil and gas extraction sho uld be sim ilar to tho se that occur in traditio nal oil and gas extraction in the petroleum industry.

The prospects for generating th e power required by the Shell *in situ* approach present some interesting challenges. For example, coal currently provides about half of the U.S. electricity supply. Thus, if coal were used to generate the el ectricity for the Shell process it would require triplin g the current rate of coal pr oduction in the United States. The impact on the U.S. coa l supply would be co mparable to that of the CTL appr oach discussed earlier. Also, the cap ital costs associ ated with the substantial increase in coal mining would need to be included in a pr oper capital cost estimate for an oil s hale industry based on this approach.

As another alternative, one c ould consider a nuclear reacto r approach to generating the electricity. This would involve buil ding about 300 one gigawatt electric power nuclear reactors.

Another approach would be to m ake use of the gas that is generated by the *in situ* process itself. Approxi mately one-third of the energy returned is contained in that gas. Thus, its energy content would be produced at a rate of about 3.4×10^{11} Watts. If that power could be converted into electricity with 50% efficiency, it could provide about half the required electricity. Sin ce the ca pital cost to construct a natural gas power plan t is about half that required to cons truct a coal fired plant, the partial production of electricity from the *in situ*-produced gas would result in about a 25% capital cost saving for the electricity generation.

It is interesting to note that the *in situ*-produced gas contains about the sam e energy as that required by the *in situ* process. If gas burners could be inserted into the shale and the retrieved *in situ*-produced gas could be pum ped into those burners and com busted, it could, in princip le, provide the en ergy needed to power the *in situ* process thereby eliminating the cap ital costs ass ociated with elec tricity generation. However, this approach, while en ergetically attractive, would likely greatly com plicate the Shell concept. It would also elim inate a potenti ally valuable/marketable product of the *in situ* process.

Shell states that the energy returned (ER) by its process (oil and gas) divided by the energy invested (EI) to obtain the energy returned is 3. This should not be confused with the process efficiency calcu lated for the o ther processes considered. That efficiency would be the energy in the oil obtained divi ded by the sum of the calorific value of the kerogen that was pyrolyzed (35 MJ/kg times the number of kg pyrolyzed) plus the energy invested to accomplish that pyrolysis. We do not know from the Shell number how much of the energy in the kerogen was captured. However, we can put an upper lim it on the process efficiency by assuming that ER represents all of the energy in the kerogen. S ince Shell states that one-third of ER was in the gas product, we will assume that two-thirds of ER is in th e oil produ ct. In that case, it is a sim ple matter to show that the p rocess efficiency would be 50%. This represents an upper lim it on the process efficiency, because it is unlikely that all of the original kerogen will be captured. For example, if half of the kerogen is captured then the process e fficiency would be about 25%. More data is needed to assess the efficiency of the Shell process.

11. Conclusion

A review is undertaken of se veral approaches to producing alternative transportation fuels using feedstocks that are under the control of the United States. The objective of the review is to provide the non-specialist reader with a general understanding of the several approaches, how they com pare regarding pr ocess energy efficiency, their indiv idual abilities to provide f or national transportation fuel needs, and their as sociated capital costs.

It was found that, for all the propulsion pl ants considered, the specific energy of the plant decreases as the sp ecific power of the plant increases. Batteries and fuel cells were shown to have a rapid decline in specific en ergy as specific power increased. As a result, there has been a very lim ited overlap between battery and fuel cel l power and energy characteristics and transportation vehicle mission requirements. In some cases, where the mission requires high power for only a sm all fraction of a vehicle's m ission, the introduction of an auxiliary pulse power system becomes viable. As a result, it is possible that battery and fuel cell technology com bined with electric motors may progress to the point where this approach is viable for pass enger vehicles where hi gh power is typically required for less than 10% of a vehicle's m ission. This could be an im portant development, because passenger vehicles account for about half of U.S. oil con sumption. However, for high-energy, high-power missions, batteries and fuel cells are disadvantaged relative to the in ternal combustion engines considered. It seem s unlikely that the disadvantage can be overco me for high-performance DOD, co mmercial, and industrial missions involving tr ansportation fuels. As a re sult, internal com bustion propulsion plants will likely remain the plants of choice for these missions.

The U.S. transportation sector consu mes about 13.4 million barrels per day (BPD) of oil equivalent product. Consequently, this is the goal that was established for each alternative fuel process considered. The alternative fuel ap proaches considered include: hydrogen production, creation of synthesis gase s from various feedstocks followed by a fuel synthesis process, enzym atic production of ethanol, the use of bio oils for biodiesel production, fuel synthesis using atmospheri c carbon dioxide as a feedstock, and the exploitation of oil shale.

A quick summary of t he findings for the va rious approaches considered is found in table 12. This table provides rough estimates of the process e fficiencies and capital costs associated with production of hydrogen and the liquid fuels cons idered at a scale needed to produce 13.4×10^6 BPD oil equivalent product. The table is ordered by increasing capital costs, except for oil shale, which is qualitatively different from the other entries.

Process	Process energy efficiency	ROM of capital cost (billions of dollars)	Comment
SMR to hydrogen	70%	173	Commercial process,* doubles NG consumption
Biodiesel	35%	200	Commercial process***
SMR to methanol	60%	250	Commercial process*
SMR to gasoline via methanol	54%	280	Commercial process*
Corn ethanol	46%	370	Commercial process***
Hydrogen by biomass gasification	46%	515	Commercial process**
Hydrogen by coal gasification	44%	525	Commercial process*, doubles coal consumption
Coal to liquid	44%	900	Commercial process*, quadruples coal consumption
Biomass to liquid	47%	900	commercial processes**
Hydrogen by thermochemical	50%	1,080	Process under development*
Lignocellulose ethanol	41%	1,467	Process under development**
Hydrogen by conventional electrolysis	25%	2,000	Commercial electrolyzer, 3'rd gen nuclear reactor*
Atmospheric CO_2	18%	4,000	Commercial processes + 3'rd gen nuclear reactor*
Oil shale surface retort	?	1,000	Involves massive mining and disposal
Shell oil shale *in situ* retort approach	Less than 50%	Greater than 500	electric power plant only*

Table 12. Summary estimates of the process efficiencies and capital costs associated with production of hydrogen and several liquid fuels at a scale to produce 13.4×10^6 BPD oil equivalent product.

Key: *not renewable but can, in pri nciple, meet the BPD goal, **renewable but available feedstock cannot sustainably meet BPD goal, ***renewable but available feed stock cannot meet BPD goal).

Several of the processes shown in table 12 can be eliminated. For example, corn-based ethanol and biodiesel have been included s imply to show where they f all in the hierarchies of efficiency and cost. In reality, while both fuels m ay have important niche roles to play, neither of them is a serious candidate for solving the national transportation fuels problem. The nece ssary feedstocks are simply not available. It is also suggested to set hydrogen aside, even though it has the lowest capital costs when produced by SMR techniques. It could be used as a fuel if it w ere absolutely necessary and could be produced in adequate quantity to meet national transportation fuel needs. If produced by nuclear or solar-powered thermochemical means or electrolysis it would produce no CO_2. However, the logistical problems associated with the cryogenic systems or high-pressure systems required to employ hydrogen as a gene ral purpose transportation fuel m ake its use as a general purpose transportation fu el or as a fuel for m ost DOD vehicles problematic.

Among the remaining processes, the steam reforming of methane (SMR) processes are found to be the m ost energy efficient and to ha ve the lowest capital costs. They can also produce a variety of different fuels, in cluding hydrogen, alcohol s, and hydrocarbons. Their use would double the consumption of natural gas and add substantially to CO_2 production. With regard to the in creased consumption of natural gas, recent adv ances in the extraction of gas from shale using hydrauli c fracture and the vast reserves of gas hydrates may play a role. The advances in extraction by hydraulic fracture have been estimated to increases the potentially available natural gas reserves by one-third, resulting in a U.S. natural gas supply of about 100 year s at current usage rates. If gas hydrates could be safely and economically obtained, they would potentially provide a large supply of methane, thereby allowing the manufacture of transporta tion fuels for generations. It was found that, while S MR processes could supply the needed transportation fuels f or an extended period, a number of issues m ust be resolved before the SM R approach to alternative fuel production can be properly assessed.

The next processes found to be most effici ent and least capital intensive were the conversion of coal to liquid fuels (CTL) and the conversion of biomass to liquid (BTL) These process are considerably m ore costly than the SMR processe s. Nevertheless, they can produce a variety of fuels, including hydrogen, alcohols, and hydrocarbons. Their use would have significant im pact on the coal and biomass resources of the United S tates. The CTL process could supply the needed tr ansportation fuels for an extended period. The BTL process could produce about 30% of the needed transportation fuel in a sustainable fashion. The BTL process woul d be carbon neutral provided that no fossil fuel was used in growing and harvesting th e biomass or in the BTL process its elf. A number of issues must be resolved before either the CTL or BTL approach to alternative fuel production can be properly assessed.

The next process to appear in the efficiency and capital cost sorting is lignocellu losic ethanol production. This process attempts to break down the cellulo se and hemicellulose in biomass into fermentable sugars. The process is still under developm ent and is much more difficult and ene rgy intensive than the well-developed corn ethanol process. However, if successful, it has the advantage th at it can acce ss a m uch larger feedstock than can the corn ethanol process. In order to function in a sustainable f ashion, it would target the sam e 1.4 billion tons of biom ass as does the BTL process. This am ount of biomass used as feedstock for the lignocellulosic ethanol process would yield about 30% of the ethanol required for a full solution. The capital investment required is estimated to be about 500 billion dollars. The situation rega rding carbon neutrality is the sam e as for the BTL approach. The process has yet to   be demonstrated at a scale where its contribution to alternative fuels can be properly assessed.

It was found that the fuel synthesis schem es with the lowest energy efficiency and the highest capital cost involve the use of atmospheric CO_2 as a feedstock. In principle, the process should be capable of producing th   e quantities of fuel needed to so lve the transportation fuel problem. For the case cons idered herein, this would require about 1600 nuclear reactors, each providing a therm al power of about three gigawatts. The associated uranium consumption would be subs tantial and would deplete the reserves of high-grade uranium ore in a few decades. Th e long-term viability of such a process would likely involve the intro duction of advanced reactors including breeder reactors . Similar approaches could be applied to converting sm okestack CO_2 to fuel, where the higher CO_2 density would m ake the collection of CO $_2$ easier. Howeve r, for the same amount of fuel the plant capit al costs would be sim ilar to that given above for fuel production from atmospheric CO_2. There are clearly many issues that need to be resolved regarding this approach to alternative transportation fuels.

The final topic considered was oil s hale. It is es timated that U.S. oil shale form ations could supply current U.S. fuel needs for m ore than 100 years. Studies done in the 1980s, when scaled to 2005 dollars, suggest that the capital cost s associated with producing 13.4×10^6 BPD of oil shale crud e by surface retorting would be a bout one trillion dollars. Shell Oil has been studying an *in situ* retorting approach in which the heat neede d to drive out the oil shale crude is provided by   electric heaters plac ed within the shale deposits. It was shown that the electrical   power system needed to provide the heat necessary to produce 13.4×10^6 BPD of oil shale crude would itself have a capital cost in excess of 500 billion dollars. There are many environmental concerns (e.g., ground water contamination) associated with producing oil shale crude.

The above results suggest these conclusions about alternative fuels:

- If necessary, the United States can manufacture the transportation fuels it needs.

- The capital investments needed to manufacture fuels beyond petroleum will be substantial, rega rdless of the particu lar alternative fuel selected. In this regard, the SMR processes, because of their high er efficiencies and substantially lower capital costs, would seem to warrant spe cial attention. Of course, the associated fuels are not carbon free or carbon neutral.

- The capital investm ents associated w ith the m anufacture of carbon free or carbon neutral fuels will be especially large. Associated with the latter point is the reality that serious investment in alternative fuels will be difficult to obtain as long as low-cost petroleum is available.

- On the basis of national security   needs, the DOD could argue to use appropriated funds to pay for the developm ent of an alternative fuel to supply its 2% of national trans portation fuel usage. Such an undertaking should be approached with great caution. If DOD se lects a scheme that is not viable for the larger transportation system , then DOD will be left with a costly proprietary system, will be unable to benef it from competitive forces in the larger marketplace, and could find itsel f short of fuel in a tim e of national emergency.

It is certainly not clear at this tim e which is the best alternative fuel ap proach for DOD and for the nation. It will like ly take decades to sort th is out. DOD should be a participant in a national e ffort to clarify the choices from a perspective of mission requirements, to ensure that these will be met, because it could be impacted substantially by the outcome.

Appendix A

A Simple Model for Vehicle Kinematics

Figure A1 provides a simple illustration of how the energy flows through a typical land vehicle to provide the force needed to drive the vehicle.

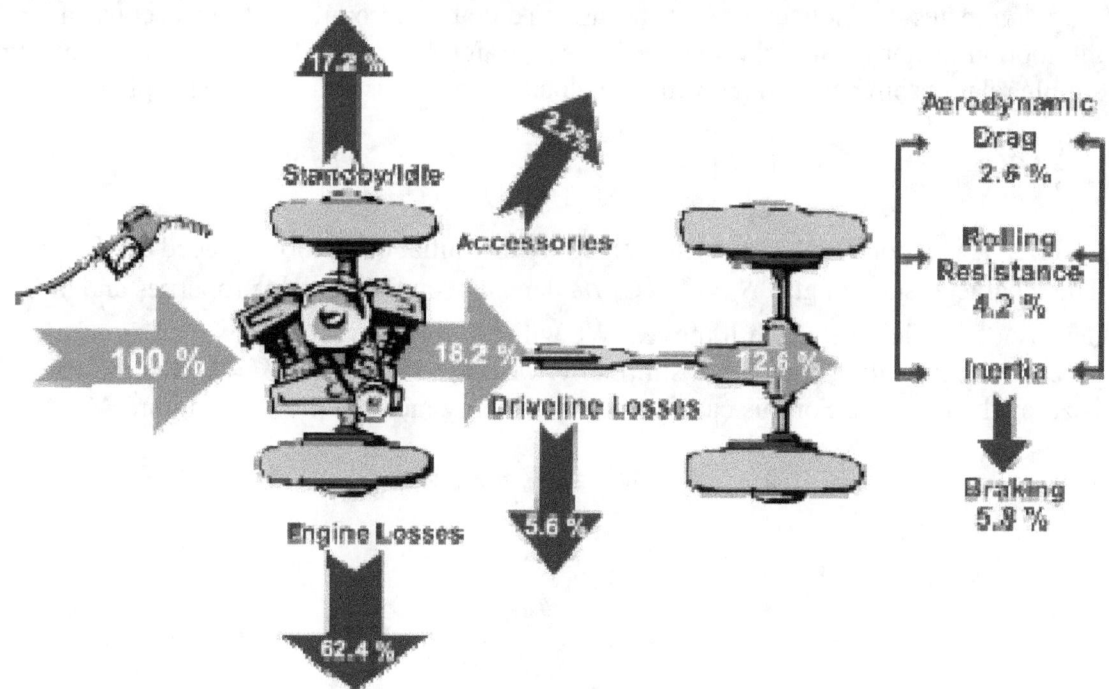

Figure A1. A Simplified Accounting of Energy Flow in a Typical Vehicle Powered by an Internal Combustion Engine. (Source: http://www.fueleconomy.gov/feg/atv.shtml)

It is obvious from figure A1 that the energy actually needed to accomplish the vehicle's mission is a small fraction of the energy consumed by the vehicle. The power plant and its management are the principal reasons for this. If one could accomplish a significant improvement on the net energy conversion efficiency of the power plant it would have a great impact on the alternative fuels problem and the fuels problem in general. Furthermore, the ultimate energy and power needs are determined by a vehicle's mission rather than the fuel or the power plant. Therefore, it is helpful to be able to make simple estimates regarding how mission requirements relate to fuel requirements. An examination of figure A1 suggests that the motion of the vehicle along a straight line can be described by the following simple differential equation:

$$M\frac{dV}{dt} = MV\frac{dV}{dS} = F_P - F_R - F_D - F_G. \qquad (A1)$$

Here M is the vehicle mass in kg, V is the vehicle speed in meters per second, t is the time in seconds, S is distance traveled in meters, F_P is the propulsion force in Newton, F_R is the rolling force, F_D is the drag force, and F_G is the component of the gravity force in the direction of motion. The rolling force can be expressed approximately as,

$F_R = C_R W$, where C_R is the rolling c oefficient and W is the vehicle weigh t. The drag force can be written as $F_D = C_D \rho A V^2 / 2$, where C_D is the drag co efficient, ρ is the fluid density, A is the appropriate area of the v ehicle (frontal area for land vehicles, surface area for aircraft, wetted area for ships and submarines), and V is the velocity of the vehicle. The quantities C_R, W, C_D, and A are properties of th e vehicle. For air vehicles, the drag force $F_D = C_D \rho A V^2 / 2$ applies only at velocities where the parasitic drag dominates the induced drag. If we assum e that the propulsion force is constant and the motion is along a straight line, then it is straightforward to show that the following simple relationship between the vehicle velocity and the distance traveled applies:

$$\frac{1}{2} M V^2 = (F_P - F_G - C_R W) S_c \left[1 - K e^{-S/S_c} \right]. \qquad (A2)$$

Here, K is a c onstant that is dete rmined by the initial cond itions placed on V and S. The characteristic length $S_c = M / C_D \rho A$ depends on the vehicle properties and the density of the fluid in which it moves. We will call the expression in square brackets the kinetic energy scaling factor. For simplicity, we will assume that $K = 1$ (i.e., the velocity is zero when $S = 0$). For this case, the kinetic energy factor is plotted in figure A2.

Kinetic Energy Scaling Factor

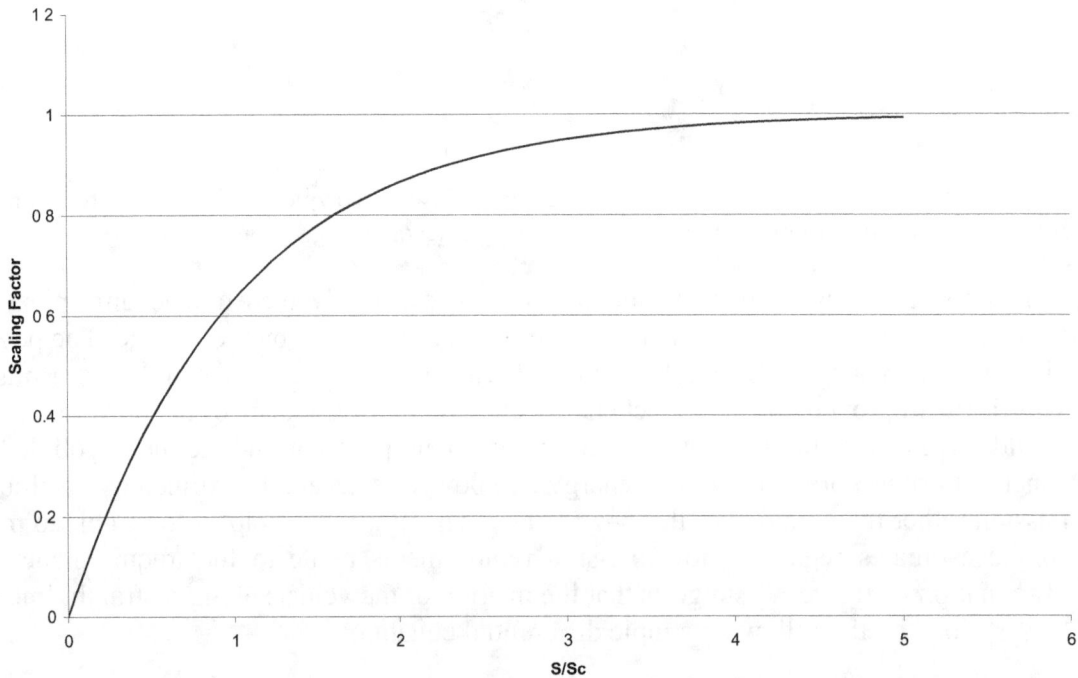

Figure A2. Kinetic energy scaling factor when $K = 1$.

Several things are ev ident from eq. (A 1) and figure A2. First, for there to be any forward motion, the propulsion force must exceed the sum of the gr avitational force and the rolling force. Second, for a fixed propulsion force, there is a m aximum velocity that

60

can be achieved. This maximum velocity is achieved when S is much greater than S_c. This region generally establishe s the power required for constant velocity travel by land or by sea vehicles. A sim ilar result holds f or aircraft, but the calculation is more complicated due to the presence of induced drag. Finally, when S is much less than S_c, the kinetic energy is proportional to S. This is a region of constant acceleration. For land vehicles, this region usually establishes the size of the propulsion plant.

In order to m ake use of eq. (A1), it is n ecessary to specify the various param eters on which the equation depends. For land vehicl es, table A1 provides som e approximate values for the param eters of several vehicles . The vehicles have b een selected to cover the range from passenger vehicles to heavy armored vehicles.

Vehicle	Weight (lbs)	Mass (kg)	Available Horse-power	C_R (asphalt surface)	C_D	A (m/s)	S_c (m)	F_R (N)
Tesla Roadster	2723	1237	288	.03	.35	2	1473	364
Honda Accord	3300	1500	270	.03	.35	2	1786	441
Jeep Grand Cherokee	4470	2032	195	.03	.45	2.8	1334	598
Hummer	6600	3000	300	.03	.6	4	1041	883
MRAP	38,000	17,000	400	.02	.8	12	1500	3,300
Abrams Tank	140,000	63,000	1500	.04	.8	12.4	5292	25,000

Table A1. Approximate Values of Parameters Characterizing Several Land Vehicles.

From table A1 and eq. A1, one can estim ate the velocities at which the drag force becomes significant relative to the rolling force. Figure A3 provides the ratio of the drag force to the rolling force as a function of speed.

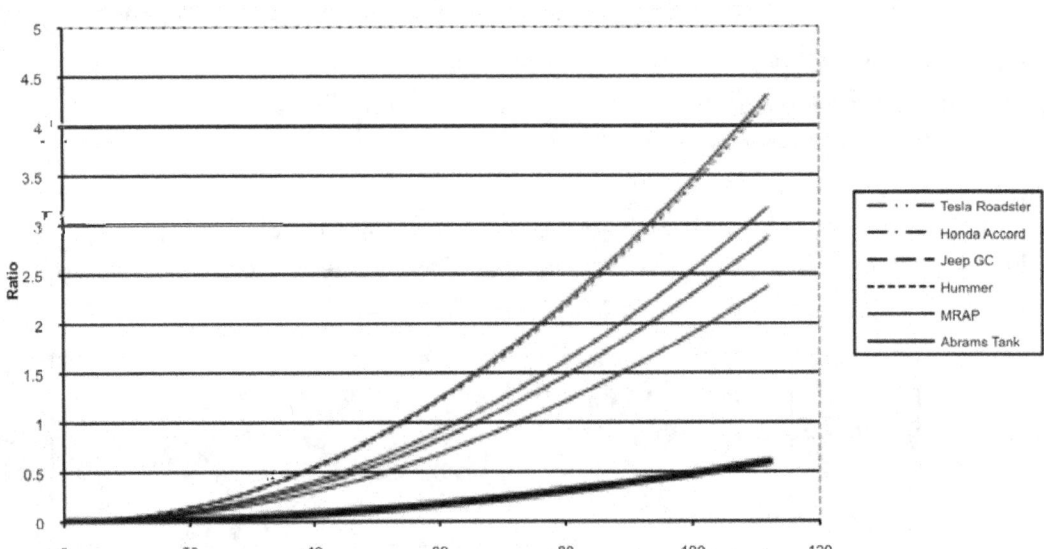

Figure A3. Ratio fo Drag Force to Rolling Force vs Speed (mph) for Several Vehicles

It is clear from figure A3 that the rolling force dominates up to speeds of about 40 mph for all the vehicles considered. One must exceed 60 mph for the drag force to become dominant. For the Abrams tank, the rolling force dominates over all speeds of interest (the Abrams has a maximum governed speed of 42 mph). The rolling force results primarily from energy lost in deforming the materials in tires, treads, and the surface on which a vehicle travels. The results presented in figure A3 are for travel over an asphalt surface. The MRAP and the Abrams have about the same drag force, but the Abrams has a much larger rolling force. This is partly due to the larger weight of the Abrams, but it is also due to the lower value of C_R for the rubber-wheeled MRAP as opposed to the tracked Abrams. Fuel consumption is related to the speed multiplied by the sum of the drag and rolling forces. One can conclude from figure A3 and table A1 that, for travel on hard surfaces, wheeled vehicles will consume less fuel per pound than will tracked vehicles. However, for travel on soft surfaces, such as sand, the value of C_R increases much more rapidly for wheeled vehicles than for tracked vehicles, and fuel consumption should be lower for tracked vehicles. This points to the obvious reality that judging fuels and fuel consumption is very dependent on mission and whether one is engaging in peacetime operations or wartime operations.

Table A1 indicates that the characteristic length S_c is measured in kilometers for the vehicles considered. In general, one wants to reach the upper speed range of a vehicle in distances less than S_c. The power that must be provided at velocity V is just VF_p. When $S \ll S_c$, the kinetic energy factor is just S/S_c. The power that must be provided at velocity V is just VF_p. It follows from eq. (A2) that, when $S \ll S_c$, the effective horsepower HP_e needed to reach a velocity V in a time t is

62

$$HP_e = 1.34 \times 10^{-3} \left(\frac{MV^2}{t} + F_R V \right). \qquad\qquad (A3)$$

The factor in front of the square brackets is the conv ersion factor from Watts to horsepower. The effective energy E_e required to accomplish the acceleration is just

$$E_e = \frac{1}{2}\left[MV^2 + S(F_R + F_G) \right], \qquad\qquad (A4)$$

where $S = Vt/2$ is the distance traveled during acceleration. Th e energy E_e is expressed in Joules. Equations (A3) and (A4) can be used to calculate HP_e and E_e if V and t are s pecified or to calculate t and E_e if V and HP_e are specified. These calculations may be done with a hand calculator.

It should be noted from eq. (A4) th at the en ergy expended to overcom e the ro lling force is pro portional to t. Hence, a s horter acceleration ti me results in less energy loss. However, the power given by eq. (A3) cont ains a term that is proportional to t^{-1} leading to the result that faster acceleration require s more power and hence a larger power p lant in the vehicle. The trad eoff between these competing effects b ecomes significant for heavy military vehicles.

If one applies eq. (A3) to the vehicles list ed in table A1 and calculates the horsepower needed to accelerate these vehicles to 60 m ph in 6 seconds and the tim e to accelerate to 60 mph with the horsepower available to each vehicle, one obtains table A2.

Vehicle	Horsepower needed to go from 0 to 60 mph in 6 seconds	Time (s) to go from 0 to 60 mph with available horsepower
Tesla	215	4.4
Honda Accord	252	5.76
Jeep GC	363	11.5
Hummer	520	12.5
MRAP	2,887	67
Abrams	11,661	103

Table A2. Horsepower needed to go from 0 to 60 mph in 6 seconds and time to reach 60 mph with available horsepower for acceleration over level ground.

Only the Tesla Roadster and the Honda Accord have the horsepower required to travel from 0 to 60 m ph in 6 seconds. The other vehicles are underpowered to m eet this objective. The third column in table A2 provi des an estimate of the time it would take to reach 60 mph with the horsepower available to the vario us vehicles "as-built." As mentioned above, the A brams tank, while it has the horsepower to reach 60 mph, is not permitted to travel faster than 42 mph.

When $S/S_c \gg 1$, It is ev ident from fig A2 that th e kinetic en ergy, and h ence the velocity, reaches a constant value independent of distance. This region determ ines the horsepower needed to maintain a constan t speed. In this case the ho rsepower can be written as

$$HP_e = 1.34 \times 10^{-3} \left[\frac{1}{2} \frac{MV^3}{S_c} + (F_G + C_R W)V \right]. \qquad (A5)$$

Fig A4 plots the fraction of the available ho rsepower needed to m aintain speed versus the speed for the several vehicles listed in table A1.

Figure A4. Fraction of Available Horsepower Needed to Maintain Speed vs Speed (mph) for Several Vehicles

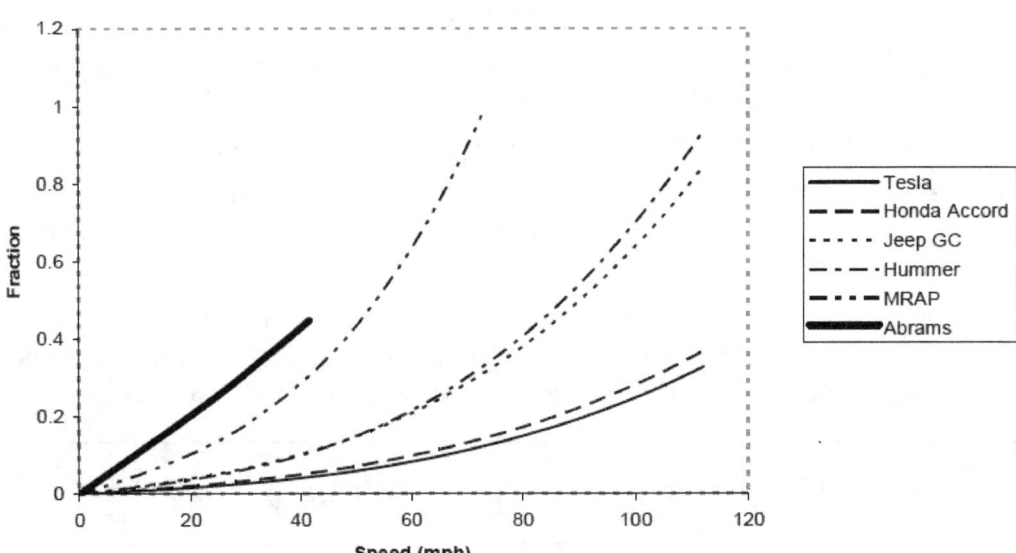

It is c lear from figure A4 that, w ith the exception of the MRAP, the horsepower available to each vehicle is larg er than that needed to maintain a particular speed. S ince the MRAP has a m aximum speed requirement of about 70 m ph, the full horsepower is required to m aintain that speed. F or most vehicles, the engines are oversized for the cruising requirement. For example, the average speed traveled by priv ate automobiles is about 45 mph. At that cruising speed, the Honda Accord uses about 6% of its available horsepower. The power plants for th ese vehicles are set by the acceleration requirements expressed through eq. (A3) and not by the cruising speed requirements.

Equation A1 can shed som e light on the energy and power requirements associated with seagoing vehicles. To illustrate, we will consider surface vess els with displacement hulls. In this case, the rolling force is zero and, for our purpos es, the gravity force can be taken as zero. The density that ente rs the characteristic leng th is $\rho = 1000$, and the area A is the wetted area of the vessel. For analytical simplicity we will use a formula for A developed by David Taylor:

$$A = 2.6(\Delta L)^{\frac{1}{2}}. \qquad (A6)$$

Here Δ is the ship displacement in tons and L is the ship waterline length in meters. The length L will be taken as 85% of the ship length. The Taylor formula is limited in accuracy but will suffice for the estimates desired here. It would be helpful to have a simple analytical relationship between ship length l and its displacement Δ. In this regard, an examination of the displacement and length data for a variety of cargo ships and fast ships finds that the data is well encompassed by two curves of the form

$$\Delta = \alpha l^3 . \qquad (A7)$$

The choice $\alpha = .0098$ is representative of the larger cargo ships. The choice $\alpha = .002$ is representative of the fast ships, such as combatants and fast cruise ships. We will refer to these as fast ships. If we use this relationship between Δ and l, then the Taylor formula becomes

$$A = 2.4\alpha^{\frac{1}{2}}l^2 . \qquad (A8)$$

.The drag force becomes

$$F_D = 12000 C_D \Delta^{\frac{2}{3}} / \alpha^{\frac{1}{6}} . \qquad (A9)$$

For cargo ships we will set $C_D = 4 \times 10^{-3}$. This value is taken from Lin et al.[3] The effective horsepower needed to maintain a ship at speed V is just the product of F_d and V. Using this data, figure 4 provides estimates for the required effective horsepower vs. desired maximum speed for cargo ships in the length range 150–400 meters

As a point of comparison, the VLCC tanker *Frank A Shrontz* with a length of about 330 meters and a cruising speed of 16 knots has a power plant of about 34,000 horsepower. This is in reasonable agreement with the predicted value obtained from figure 4 and suggests that the power plants chosen for cargo ships are selected to be close to the power needed to maintain a desired cruising speed. Similar calculations can be done for fast ships when the appropriate values of the drag coefficient and wetted surface areas are known.

Appendix B

The Gasification and Fischer-Tropsch Processes

For the purpose of illustration, this appendix considers a specific example of a Fischer-Tropsch reactor fed by a coal gasifier. The basic FT synthesis process proceeds according to the reactions

$$nCO + 2nH_2 \rightarrow C_nH_{2n} + nH_2O \qquad (B1)$$

$$nCO + (2n+1)H_2 \rightarrow C_nH_{2n+2} + nH_2O \qquad (B2)$$

The first reaction produces olefins, or alkenes, (C_nH_{2n}) while the second reaction produces paraffins, or alkanes, (C_nH_{2n+2}). As an example, when n = 8, the reader will recognize the paraffin product as octane. The FT process always produces a mix of olefins and paraffins. However, the details of the mix depend on the reactor conditions and the catalyst used. The FT reactors are generally operated in a low temperature range of 200–240 $^{\circ}C$ or a high temperature range of 300–350 $^{\circ}C$. Low-temperature operation favors high molecular weight waxes, while high-temperature operation favors low molecular weight olefins. The route to diesel fuel is through the low-temperature process, while the route to gasoline is the high-temperature process. In both cases, product upgrading is required.

A typical source of the carbon monoxide (CO) and hydrogen needed by the FT reactor is a gasifier. Because of the central roles that gasification and Fischer-Tropsch chemistry play in the study of alternative fuels, this topic is discussed here in somewhat more detail than others in this paper. The discussion will still be elementary. A more thorough discussion of gasification can be found in Probstein and Hicks.[2]

For this example, consider a gasifier that can produce a syngas with a H_2/CO ratio of 2:1, which it achieves by converting a substantial fraction of the available CO to CO$_2$. Although the ratio of 2:1 is necessary for the olefins FT process, this greatly reduces the total quantity of CO in the feedstock. There is, therefore, a tradeoff between achieving the desired ratio and the absolute amount of CO available for the FT process. To the extent that the H_2/CO ratio is less than 2:1 it will be necessary either to burn additional coal or to provide supplemental hydrogen in order to produce the needed quantity of desired product.

The actual amount of CO and hydrogen available is determined by the chemical kinetics that occur in the gasifier. For the application considered here, a gasifier works by combusting a fraction of the available carbon and using the released energy to raise and heat steam and carbon to a temperature where the production of CO and H_2 is optimized. In this regard, a temperature of 1000 $^{\circ}K$ is appropriate. In figure B1 we provide an oversimplified illustration based on equilibrium chemistry for the case where one mole (12 grams) of carbon is gasified using .25 mole (8 grams) of oxygen and 1 mole (18 grams) of water.

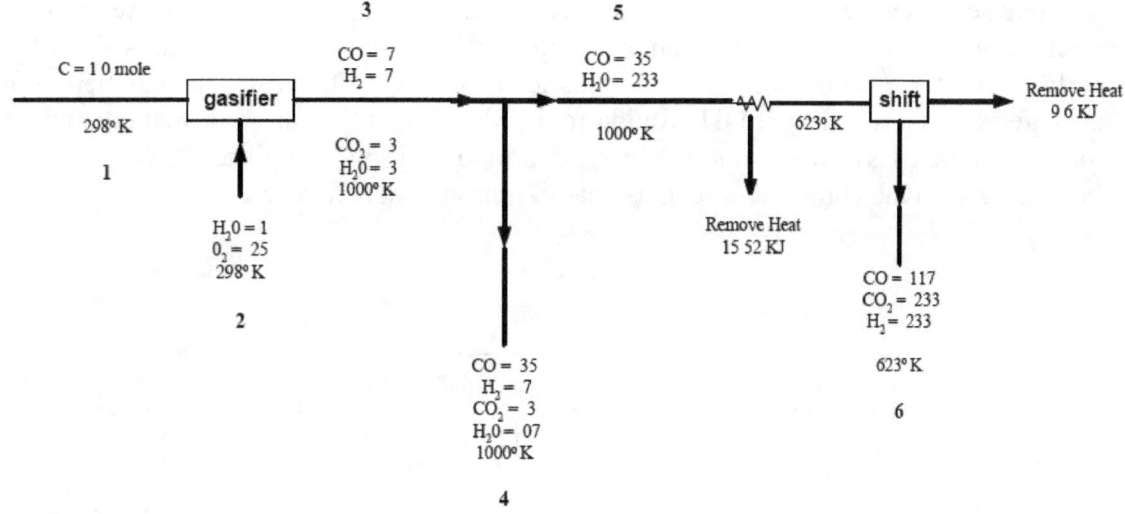

Figure B1. Mass flowchart (units in m oles) from the gasifier through the water-gas shifter.

For the m oment, we will consider only str eams 1, 2 and 3 of figure B1. The combustion of .25 m ole of carbon with .25 mole of oxygen produces 98.6 KJ of heat energy. This is adequate to raise and heat the 1 mole of steam and the remaining .75 mole of carbon to 1000 °K. In this particular ex ample, the gasification reactions that lead to stream 3 will not greatly alter the energy balance. In general, if the gasification reactions lead to an energy def icit regarding maintaining the desired temperature, then additional energy would have to be added. Similarly, if the gasification reactions produced energy excess to m aintaining the desired tem perature, then energy would have to be removed. Excess energy could, in principle, be reused, for exam ple, to raise s team or to generate electricity.

The gasifier outlet stream 3 contains only half the hydrogen needed to satisfy equations (B1) and (B2). This m eans that only one half of the carbon in stream 3 can be converted via the FT process. T his is very wastef ul in the use of carbon (e.g. coal). Som e improvement can be made by introducing an additional step that employs the w ater-gas shift reaction

$$CO + H_2O = CO_2 + H_2 \qquad (-41.2 \text{ kJ/mol}) \quad (B3).$$

Figure B1 illustrates the use of the shift r eaction by separating stream 3 into stream 4 and stream 5. Stream 4 contains all of the hydr ogen from the gasifier and half the carbon monoxide. The remaining carbon monoxide and suffi cient steam is separated into strea m 5 such tha t the outpu t of the shif t reactor (stream 6) w ill contain the proper ratio of hydrogen to carbon monoxide. The catalyst in the shift reactor works best at 623 °K. This requires that heat (15.5 KJ) be removed from stream 5. Since the shift reaction is exothermic, heat (9.6 KJ) m ust also be rem oved from the shift reactor in order to maintain its temperature. The excess heat from stream 5 and from the shift reactor can, in principle, be reused.

If one separates out and combines the carbon monoxide and hydrogen in streams 4 and 6, one is left with a combined stream containing .933 mole of hydrogen and .467 mole of carbon monoxide. This approximately satisfies equation 2. Therefore, the gasifier-shifter arrangement shown in figure B1 would provi de 47% of the car bon that entered the gasifier to the FT synthesis unit. F igure B2 provides an oversimplified mass flow chart beginning with the combined stream (stream 7) through the FT reactor.

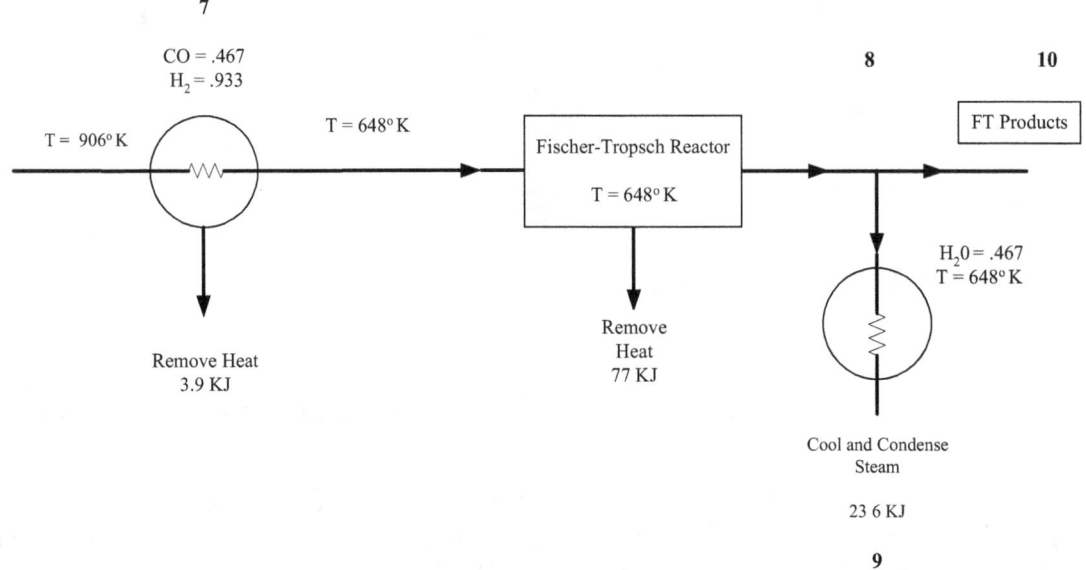

Figure B2. Mass flow chart (flow units in moles) from the water-gas shifter through the FT reactor.

The FT reactor prefers to operate at a temperature of 648 °K. This requires that 3.86 kJ of heat be removed from stream 7. Furthermore, the FT reaction is exothermic. In order to maintain the FT reactor temperature at 648 °K it is necessary to remove 77 kJ of heat from the reactor. The FT reactor is the largest source of excess heat in the example considered here.

The FT reactor generally produces a mixture of compounds. The liquid fuel fraction has a maximum selectivity of about 50%. At this selectivity, about 25% of the carbon that entered the gasifier would exit the FT synthesis unit as gasoline or diesel. This implies that one ton of carbon (coal) produces about .29 tons, or 2 barrels, of oil equivalent liquid. This is in agreement with industrial experience with the coal to liquid (CTL) process Therefore, about 2.2×10^9 tons/year of coal would be needed to provide the liquid fuel to replace conventional petroleum using the CTL process. From an energy efficiency perspective, it should be noted that one ton of coal has a calorific value of about 28 GJ while two barrels of gasoline have a calorific value of about 12.2 GJ. This results in a net calorific efficiency of about 44% for the CTL process. There is the potential to improve this efficiency by utilizing some of the waste heat that is evident in figures B1 and B2. The excess heat identified in these figures is about 130 kJ. If 50% of this energy could be recovered and used to generate electricity, then the net calorific efficiency would rise to about 57%.

One can gain some understanding of the total power flow that would be involved in a CTL solution of the national transportation fuels problem by scaling from the 12 grams of carbon used in the example above to the needed 2.2×10^9 tons/year (6.3×10^7 gm/sec). In this case, the thermal power associated with carbon combustion is 5.2×10^{11} Watts. This corresponds to 170 $3GW_{th}$ coal gasifiers. The thermal power associated with the waste heat identified in this simple example would be about 7×10^{11} Watts. Since the average U.S. electrical power usage is about 5×10^{11} Watts, there would be a high priority placed on making use of the waste heat from CTL plants associated with a solution to the national transportation fuels problem.

References

1. EIA Annual Energy Outlook 2006, DOE/EIA-0383(2006).

2. Synthetic Fuels, R. Probstein and R. Hicks, Dover (2006).

3. C. Lin, S. Percival, and E Gotimer, "Viscous drag calculations for ship hull geometry" http://www.dt.navy.mil/hyd/tec-rep/vis-dra-cal/documents/viscous.pdf

4. D. Ragone, Proc. Soc. Automotive Engineers Conference, Detroit, MI., May (1968).

5. D. Georgi, "Lithium Primary Continues to Evolve," Batteries Digest: http://www.batteriesdigest.com/lithium_air.htm.

6. The Cunard Liner Queen Elizabeth 2: http://www.qe2.org.uk/engine.html.

7. T. Coffey, D. Hardy, G. Besenbruch, K. Shultz, L. Brown, and J. Dahlburg, "Hydrogen as a fuel for DOD," Defense Horizon number 36, National Defense University.

8. D. Simbeck and E. Chang, "Hydrogen Supply: Cost Estimate for Hydrogen Pathways-Scoping Analysis," National Renewable Energy Laboratory, NREL/SR-540-32525 November 2002.

9. K. R. Schultz, L. C. Brown, G. E. Besenbruch, and C. J. Hamilton, "Large-scale production of hydrogen by nuclear energy for the hydrogen economy," GA-A24265 (https://fusion.gat.com/pubs-ext/MISCONF03/A24265.pdf) , February 2003.

10. M. Mintz, J. Molburg, S. Folga, and J. Gillette, "Hydrogen Distribution Infrastructure," Hydrogen in Materials and Vacuum Systems: First International Workshop on Hydrogen in Materials and Vacuum Systems. AIP Conference Proceedings, Volume 671, pp. 119–132 (2003).

11. P. Spath, A. Aden, T. Eggeman, M. Ringer, B. Wallace, and J. Jechura, "Biomass to Hydrogen Production Detailed Design and Economics Utilizing the Battelle Columbus Laboratory Indirectly-Heated Gasifier" NREL TP-510-37408, May 2005.

12. T. L. Buchanan, M. G. Klett, and R. L. Schoff, "Capital and Operating Cost of Hydrogen Production from Coal Gasification," National Energy Technology Laboratory, April 2003 (http://www.fischer-tropsch.org/DOE/DOE_reports/40465/40465-FNL.%202003/40465-FNL,%202003_toc.htm).

13. J. Ivy, "Summary of Electrolytic Hydrogen Production," National Renewable Energy Laboratory NREL/MP-560-36734, September 2004.

14. J. S. Dukes, "Burning Buried Sunshine: Human Consumption of Ancient Solar Energy", Climatic Change 61:31–34, 2003 (http://globalecology.stanford.edu/DGE/Dukes/Dukes_ClimChange1.pdf)

15. M. K. Hubbert, "Nuclear Energy and the Fossil Fuels 'Drilling and Production Practice" (June 1956), http://www.hubbertpeak.com/hubbert/1956/1956.pdf.

16. http://www.mines.edu/Potential-Gas-Committee-reports-unprecedented-increase-in-magnitude-of-U.S.-natural-gas-resource-base.

17. M. Brusstar and M. Bakenhus, "Economical, High-efficiency Engine Technologies for Alcohol Fuels", http://www.methanol.org/pdf/ISAF-XV-EPA.pdf.

18. I. Wender, "Reactions of synthesis gas," *Fuel Processing Technology* 48(3): 189-297 (1996).

19. S. L Meisel,. J. P., McCullough, C. H. Lechthaler, and P. B., Weisz, "Gasoline from Methanol in One Step," ChemTech 6, 86–89, 1976.

20. J. Ansorge, "Shell Middle Distillate Synthe sis: Fischer-Tropsch Catalysis in Natural Gas Conversion to High Quality Products," available at: http://www.anl.gov/PCS/acsfuel/ preprint%20archive/Files/42_2_SAN%20FRANCISCO_04-97_0654.pdf

21. "Liquid Fuels from U.S. Coal," http://www.nma.org/pdf/liquid_coal_fuels_ 100505.pdf

22. R. D. Perlack, L. L. Wright, A. F. Turhollow, R. L. Graham, B. J. Stokes, and D. C. Erbach, "Biomass As Feedstock for a Bioenergy and Bioproducts Industry: The Technical Feasibility of a Billion-Ton Annual Supply," available at: http://www1.eere.energy.gov/ biomass/pdfs/final_billionton_vision_report2.pdf

23. R. Wallace, K. Ibsen, A. McAloon, and W. Yee, "Feasibility Study for Co-Locating and Integrating Ethanol Production Plants from Corn Starch and Lignocellulosic Feedstocks," NREL/TP-510-37092 (January 2005).

24. J. Sheehan, V. Camobreco, J. Duffield, M. Graboski, and H. Shapouri, "An Overview of Biodiesel and Petroleum Diesel Life Cycles," National Renewable Energy Laboratory report NREL/TP-580-24772, May 1998.

25. J. Sheehan, T. Dunahay, J. Benemann, P. Roessler, and J.C. Weissman, "A Look Back at the U.S. Department of Energy's Aquatic Species Program—Biodiesel from Algae," National Renewable Energy Laboratory report NREL/TP-580-24190, July 1998.

26. Stucki, S., Schuler, A., Constantinescu, M., "Coupled CO2 recovery from the atmosphere and water electrolysis: Feasibility of a new process for hydrogen storage," *Int. J. Hydrogen Energy* 1995, *20*, 653–663.

27. F. Jeffrey and W. Kubic, "Green Freedom: A Concept for Producing Carbon-Neutral Synthetic Fuels and Chemicals," Los Alamos National laboratory LA-UR-07-7897, November 2007, available at: http://www.lanl.gov/news/newsbulletin/pdf/Green_ Freedom_Overview.pdf.

28. Office of Technology Assessment, "An Assessment of Oil Shale Technologies," June 1980, available at: http://www.gwpc.org/e-library/documents/general/ An%20Assessment%20of%20Oil%20Shale%20Technologies.pdf.

29. "Survey of Energy Resources: Oil Shale," World Energy Council, December 2000, available at: http://www.energybulletin.net/print/5600.

30. Shell Oil Technology In situ Conversion Process, information available at: http://www.shell.us/home/content/usa/aboutshell/projects_locations/mahogany/ technology/